Communication for professional engineers

COMMUNICATION

for professional engineers

Bill Scott

Thomas Telford Ltd
London 1984

Published by Thomas Telford Ltd, 26–34 Old Street, London EC1
ISBN: 0 7277 0187 8
© Bill Scott, 1984

British Library Cataloguing in Publication Data

Scott, W. P.
Communication for professional engineers
1. Engineering—Management 2. Communication
in engineering 3. Communication in
management
I. Title
658.4'5'02462   TA190

ISBN 0–7277–0187–8

Set in 11/12pt Baskerville by Santype International Ltd, Salisbury, Wilts.
Printed and bound in Great Britain by Redwood Burn Limited,
Trowbridge, Wiltshire.

Dedication: SNWS

Note to reader

This book is about the skills of communication. It is in four main parts:

All levels of engineer should find it significant: the senior engineer concerned with the conduct of efficient meetings and interviews; the mid-career professional faced with the responsibility of representing himself and his organisation to the outside world; the junior engineer wanting to acquire a good background for his future career development.

The skill of communicating effectively is rarely an inherited gift. The majority of us, not blessed with that instinctive flair, can nevertheless develop the ability. It's not easy, and it needs hard work. It depends partly on acquiring an understanding of technique, such as can be gained from this book. It depends also on practical development of competence and confidence, either through on-the-job coaching by sensitive management or, more often, off-the-job in the sort of seminar in which we can rapidly build compressed experience.

The origins of the book

The author of this book has been responsible for the training in communication skills of professional staff during a period of over 20 years. He works regularly in seminars with leading professionals, usually well into their careers, both in the UK and abroad. He is retained as consultant by the Institution of Civil Engineers to run such programmes, and his private clients include engineers, scientists, accountants, economists and businessmen.

In this book he is drawing on materials which have been developed for seminars with those professionals; which have been tested with them, improved through their comments and criticisms and refined, not only in the light of his own experience, but also in the light of very positive experience which they have fed back to him.

A group of eminent engineers has advised on the drafting of the book and the author is greatly indebted to them for ensuring that the book is accurately orientated to engineering practice.

The result is a practical handbook on communication for professional engineers.

W P Scott
12 Trafalgar Road
Southport

Acknowledgements

Stella Ascott
Poul Beckmann, MSc, MIngF, MICE, FIStructE
Donald F. Dean, CBE, BSc, FICE, FIHE
Brian Derby
Jeremy Swinfen Green, MA
David F. Sutton, JP, BA
J. D. Wallace, BEng, FICE

Engineering Advisory Group:
Mrs D. M. M. Brown, BSc, MICE
Hugh Ferguson, BSc(Eng), Diploma TE, MICE, MIHE
Dick Jones, MA, MBIM
Reg Main, FIMechE
John H. Sargent, FICE, FGS
Vernon de Silva, BSc

Contents

Part 2. Effective writing

Part 1

Effective speaking

Introduction to Part 1

Part 1 of this book deals with effective speaking: the skills of winning the interest and acclaim of an audience.

The first chapter deals with the problems met by those who have to speak in public: the difficulties which they have in adjusting themselves to this challenge and the barriers which they meet when talking to an audience.

To be effective, the speaker must infect his audience with his own enthusiasm. This demands that he use his energy to project that enthusiasm, and he can do so only if he is properly prepared and has an adequate discipline with which to project his message. Preparation is the topic of Chapter 2 and a discipline which can be used regularly, mechanically, to structure a talk is outlined in Chapter 3.

Chapter 4 deals with further steps which the speaker can take to earn the enthusiasm and interest of his audience.

Visual aids can powerfully reinforce—or powerfully distract from—the effect of a talk. They are the topic for Chapter 5.

Conference speeches—the sort of speech that is supposed to last for an hour or so—are a special hazard (see Chapter 6).

The converse of effective speaking is effective listening, the topic for Chapter 7.

I

The engineer as speaker

Engineers are erudite, intelligent, experienced. They have a big vocabulary and are good at chatting informally with one another. They have knowledge and experience of fascinating projects.

But when it comes to putting these ideas over publicly, especially to a lay audience, they often feel worried. They can be seen to be uncertain, apologetic, diffident. They can be heard to be confused, difficult to understand, rambling.

This opening chapter of the book is concerned with the hazards which they face, the obstacles they have to overcome, and criteria for their success in preparing and presenting a speech.

The chapter is in three sections:

1. Comprehension: what people remember
2. Barriers to communication
3. Implications and action needs

Comprehension

Engineers are often surprised by the extent to which they can be misunderstood or even, apparently, ignored. This does not spring entirely from their own shortcomings: it is always a danger within the processes of communication between human beings.

> As an illustration: I regularly ask engineers to give short talks to half-a-dozen of their colleagues. Those col-

leagues then remember quite different interpretations of a short speech.

If each listener has heard, say, a dozen points, then there may be only one of the dozen which has been remembered by all of the other listeners.

There are never more than four points common in the recollection of all the listeners.

Always, each listener remembers at least one point which has not struck his colleagues at all.

There is, in this experiment, a dramatic difference in what is heard and remembered of the same talk by different listeners. It illustrates the normal imperfections regularly found in everyday communications.

The divergence—the difference between the sent message and the received message—is of course increasingly serious as messages get increasingly complicated. People understand clear thinking. Practical people understand simple thinking.

Let us consider some of the obstacles which create the need for communication to be simple.

Barriers to communication

There is a series of stages through which a listener comes to accept or remember a message spoken to him. Between each successive stage and the next, there are barriers to communication.

First, the listener may not hear.

He may be dozing off.

There may be outside noise.

The speaker may articulate badly.

The power of voice may be inadequate.

The microphone may distort.

There may be language and dialect problems.

Second, the listener may not understand what he hears.

Difficult words.

Difficult thought processes.

6

Poor organisation of material.

Convoluted thinking.

Educational and technical deficiencies.

Language barriers.

Jargon.

Misinterpretation, either deliberate or accidental.

Third, that which is understood may not be accepted. This may of course be for rational reasons, but there are also plenty of other hazards.

Low feeling of involvement.

Vested interests elsewhere.

Conflicting objectives.

Poor chemistry between listener and speaker.

Fourth, the speaker may lack feed-back. The inexperienced speaker, talking to an audience, is often quite unaware of what his listeners are thinking. He reads his speech or concentrates his thinking on what he is saying. He is unlikely to discover that the audience is bored unless he hears a snore. So it may be only in question time subsequently that he discovers they have completely missed the point.

The engineer talking informally with one of his colleagues, on the other hand, is constantly exchanging comments with him, constantly getting feed-back about the other's interest and understanding of what he is saying.

In public, the speaker needs some alternative means for constantly monitoring and reacting to the audience's mood.

Implications

The speaker who wants to overcome these barriers has to develop a series of skills.

To ensure that he is heard, he must project the message. He must ensure that the way in which he speaks overcomes the physical range of barriers.

To ensure that what is heard is understood, he must organise the message in a way which is simple, and he must deliver it in a way which is so pointed that the listener is helped to understand. He needs a technique for *preparation*, and a drill he can use for the *structure* of his talk.

To earn the highest chance of his message being accepted, he must establish a positive relationship with his listeners. The chemistry between speaker and listener—the *interaction*—must be good. This depends in part on the form of the message—on the preparation and structure—but more than that it depends on very positive steps being taken to project energy and enthusiasm. At its most effective such interaction also enables him to get some feed-back about the listener's response.

The next three chapters deal with these critical elements of technique in speaking: preparation, structure and personal interaction.

Summary

It is natural that much of a message may be ignored or misunderstood.

Different listeners receive different messages.

There are barriers to hearing, to understanding, to accepting.

The speaker must develop techniques for preparation, for structuring a talk, for projecting energy and enthusiasm, and for gauging and reacting to an audience's interest.

2

Preparation for speaking

The foundation for an effective speech is competent preparation.

I take it for granted that the engineer knows his subject. He has the facts, the figures, the experience which he needs. What he may lack is a technique by which he can put his ideas into the form of an effective speech.

Technique, both for preparation and for presentation of a speech, depends in part on the form of speech to be given. There is a clear distinction between trying to capture the interest of a lay audience, and the delivery of a weighty technical speech to professional colleagues.

In this chapter (and in the next two chapters), the emphasis will be on the former: the speech to interest a lay audience, and it is assumed that the speech will be of modest duration—not more than half-an-hour. Variation of technique for longer or more technical speeches will be discussed later (Chapter 6).

This chapter on preparation for speaking is in four sections:

1. Key concern
2. Possible approaches
3. Constraint
4. Recommended method

Key concern

In speaking to somebody, in asking questions of somebody, in giving information to somebody, in giving instructions to somebody:

It is the *somebody* who is of key concern. The need is to get a message over to *him*.

The need is to choose material which is appropriate for *him*.

The need is so to prepare that the speaker's energy can later be used to enthuse *him*.

Note that this is entirely different from the normal key concern of the engineer.

He is normally intrigued by the subject of engineering.

He is concerned with the precision of the subject.

He is concerned also with the imprecision of the subject: with the areas in which his experience and judgement must compensate for absence of scientific certainty.

He is concerned with the difficulty of breaking his subject down into digestible, yet cohesive sections.

He is not normally so concerned to think of the listeners and of their interests. Yet his success in communicating is not going to depend on his subject competence—we take that for granted. It will depend on his success in choosing the right message for them. He should stop worrying about what he's going to say, and start worrying about what they're going to hear.

This leads us to three criteria for the preparation of a speech.

1. The message must be so chosen that it will be of interest to the particular group of listeners.

2. It must be in a form which they will understand.

3. It must be in a form which will enable the speaker, when delivering the speech, to spread his enthusiasm to the listeners.

Possible approaches

There are a variety of approaches to preparing a speech. Let us start with two: impromptu, and extensive writing.

IMPROMPTU

This is the method adopted by the speaker who believes that every speech must be appropriate to the audience, to their mood when he starts and as his talk develops, and that he need not prepare beforehand. It is a method which assumes the heights of confidence in the speaker and which also requires an extraordinary ability to marshal and to project ideas while standing on one's feet in front of an audience.

Some small percentage of people have a natural talent in this form. It is an ability which enables them to co-ordinate several mental processes at the same time.

Such talent is an instinctive and intuitive ability. It corresponds with the natural ability of some athletes, some batsmen for example, able to play a natural game with very little coaching.

Such people are rare. In my experience the real talents needed to be an 'off-the-cuff' speaker are not found in more than about one person in a thousand.

That isolated individual has a rare talent. It is only confused by the attempt to coach or to develop more systematic methods. If you happen to be the rare speaker who has that talent—please skip the rest of this chapter.

EXTENSIVE WRITING

As a rule, it is not advisable to write a full speech. There are exceptions to the rule—they will be mentioned in Chapter 6—but in general, the method has several drawbacks.

For those who try to memorise a full script and later deliver it from memory there is need for great energy and concentration on remembering. There is at the same time the risk that the memory can fail—and when it does, it can be catastrophic.

If the script is read, it sounds as if it is being read. The message is in prose and the projection of the voice, when reading with a bent head, is away from the audience.

Whether remembered or read, the written script is subtly, but substantially, different from that which would otherwise be spoken. There are differences in choice of words, differences in phraseology, differences in inflection, differences in the structure of sentences and paragraphs. The written word, when read, therefore lacks the immediacy of the spoken word.

Equally serious, the written statement is inflexible. The speaker cannot react to the mood of the audience. He can neither cut nor amplify passages in response to their interest.

Writing is for a reader, not for a listener.

Midway between the extremes of impromptu and fully written preparation are various forms of note-making. It is one such form which will shortly be recommended, but first we must consider a constraint.

Constraint

People's capacity to put their thinking into order is of course determined by the quality of the human brain.

The brain is a remarkable tool. It has a capacity to remember, to be creative, to analyse, to articulate; but, all too often, when people are trying to put their thoughts into order, they find it very difficult. They find that despite knowing their subject matter very well, they cannot easily sort it out into a simple framework. They become frustrated, set the task aside, come back to it repeatedly. Their progress is slow and depressing.

The cause of the problem lies in the way they use their brain, and such people are in fact misusing the brain in two ways.

First, they overload. They ask the brain to produce knowledge from the memory bank. At the same time, they ask the brain to put it into some sort of order.

And even at the same time, they might try to articulate—in writing or in speech. The brain is overloaded; it then becomes incapacitated; it is then very frustrated.

Second, there is a problem of clarity of vision. People can understand and discuss issues which they see clearly. They can cope with big and complex issues, provided that they recognise those issues clearly.

But if an issue is seen imprecisely, ambiguously, murkily, then even the trivial becomes difficult.

For both the speaker and the listener the issues—however complicated—must be expressed simply and clearly.

The constraint then is set by the human brain. In preparing his speech, the speaker must avoid overloading his own brain and must analyse simply enough for his own brain and for those of his listeners.

Recommended method

The approach recommended for getting material into such simple order is a three-stage process. The stages involve using different sizes of paper: A4, A5 and A6.

The first stage is the production phase. The human mind can be extremely productive for a short period. The speaker should dig into his memory bank intensively and should quickly extract from it the range of ideas he has on his subject.

The purpose of this exercise is primarily to clear the brain so that it can later go on to its analytical function. There is therefore no need for any sort of order or accuracy of thinking at this first stage.

The method is to take an A4 sheet of paper, to set down the title at the top, then at random to set down all the ideas which come through the brain on that title.

Properly used at this stage, the brain can produce a lot of ideas very quickly: more quickly than it is possible to write out each idea. The need is quickly to set down one or two key words—not more—for each idea and then quickly to move on to the next word for the next idea.

Do not worry that some words may have little relevance— get them down or they will block the flow of thought.

If the flow happens to go right off the subject, do not worry. Look back half-a-dozen lines and pick up the thinking from that point.

So in an intense period of concentration, the A4 could well become covered with a couple of columns of ideas.

The ability of the brain to be so productive is limited. It is at its best when forced to concentrate for a short period. For one person working on his own, that period is probably not more than two minutes.

The second stage of the preparation is the analytical phase.

The need now is to plan a speech so that it will meet the two criteria:

1. The message must be chosen so that it will be of interest to the particular group of listeners.
2. It must be in a form which they will understand.

For this purpose, first turn over the A4 sheet of paper. Symbolically, switch away from the first (production) phase.

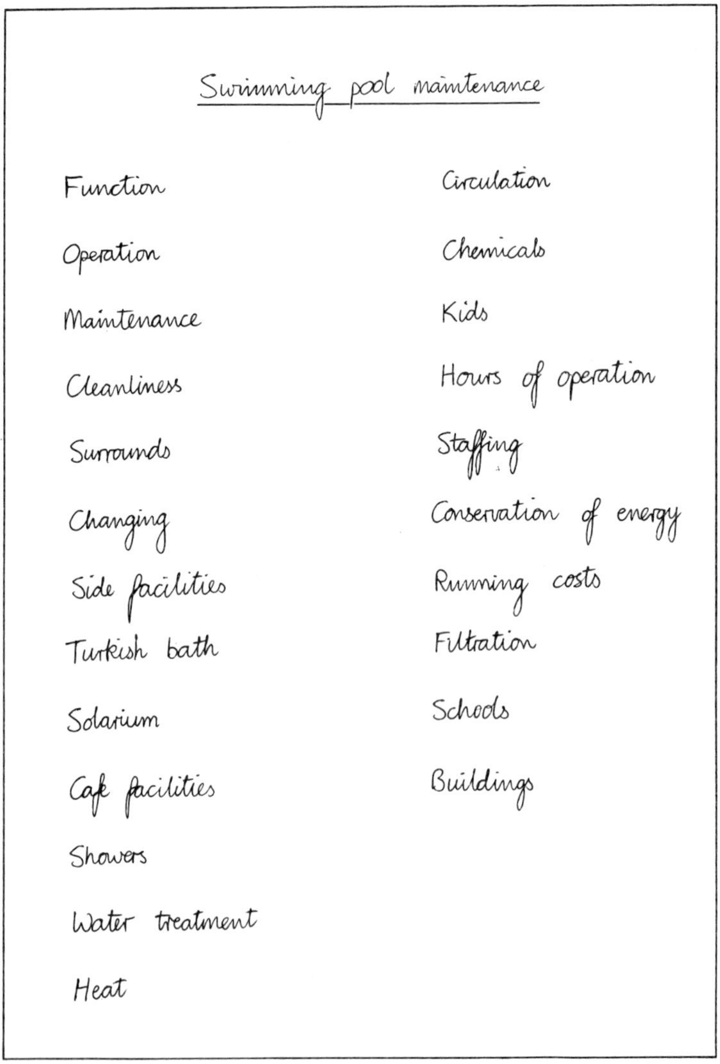
Swimming pool maintenance

Function
Operation
Maintenance
Cleanliness
Surrounds
Changing
Side facilities
Turkish bath
Solarium
Cafe facilities
Showers
Water treatment
Heat

Circulation
Chemicals
Kids
Hours of operation
Staffing
Conservation of energy
Running costs
Filtration
Schools
Buildings

Figure 2.1 Example of the first stage in preparing a talk—the A4

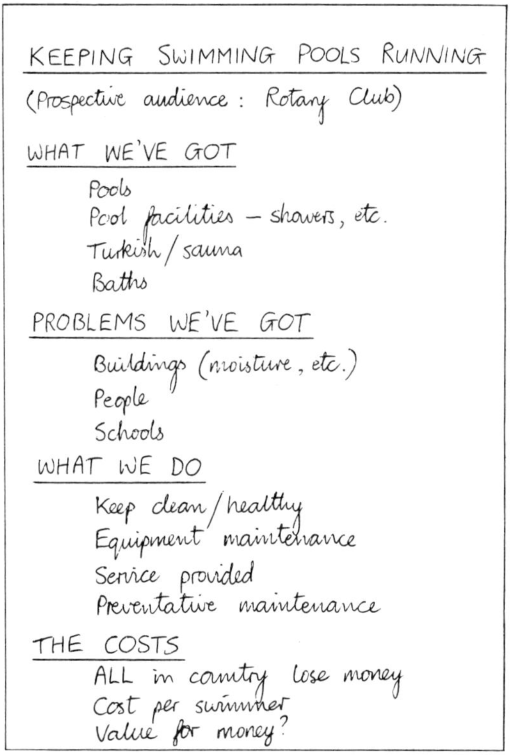

KEEPING SWIMMING POOLS RUNNING

(Prospective audience : Rotary Club)

WHAT WE'VE GOT
  Pools
  Pool facilities — showers, etc.
  Turkish / sauna
  Baths

PROBLEMS WE'VE GOT
  Buildings (moisture, etc.)
  People
  Schools

WHAT WE DO
  Keep clean / healthy
  Equipment maintenance
  Service provided
  Preventative maintenance

THE COSTS
  ALL in country lose money
  Cost per swimmer
  Value for money?

Figure 2.2 Second stage in preparing a talk—the A5

Take a clean sheet—an A5. Set down the title at the top.

Now stop and change wavelengths. Help to change the mental wavelength by changing the physical wavelength—by changing chairs or sitting well back in the present chair.

Think of the audience. Consider what sort of people they are—their education—their likely experience—their age range—their profession ... their knowledge of the subject ... what will interest them in this subject ...

> **... *What are the four headlines of most interest to them?***

Note that this thought process is now fulfilling two ends.

It is listener orientated. It is considering that which is of

Figure 2.3 Third stage of preparation—the A6

> interest to the audience. It is not based in the complexities of the content, which were turned over with the A4 sheet.

> It is simple. It is concentrating on analysis under four headlines—a number which the human brain can comfortably grasp.

> This is not to say that it is a serious error to analyse under three headings or five headings. But neither the speaker nor the listener will have an equally clear and precise understanding of a topic if it is analysed under six, seven or eight main headings.

Having on the A5 established the four main headings, the speaker goes on to complete his second stage preparation. He should write in three or four supporting sub-headings under each main heading.

It is again important to recognise the need to break down into digestible chunks. If the speaker tries to put a dozen sub-points under one of his main headings, then that will become too diffuse a section for him to communicate. Rather, he

16

should break the section into three or four secondary points and, if necessary, break those further into tertiaries.

In the second phase of his preparation, then, the speaker has taken an A5 sheet and, concentrating from his perception of the listener's interest, created a framework for his speech under four main headings.

This analytical process will have given him a clear and sharp picture of what he needs to communicate to his audience. He should be able to go on to give his speech, confident in the clarity of his thinking, able to use his energy then to enthuse the audience.

He will possibly feel the need for some sort of 'prompt'. The A5 is too big a document to serve that purpose. It has on it more than the speaker can see at one glance of the eyes and so it becomes a magnet for his interest and energy. It distracts him from his audience, makes him lose contact with them.

Phase 3 of the preparation is the simple one of taking a sheet of A6—something the size of a postcard—and printing on it not more than half-a-dozen words. Printed large. One (or two) words for the title and for each of the headings.

If the preparation has been systematically carried through the A4 and A5 phases, this simple prompt is enough to re-awaken his subconscious when he is talking.

This is a process to be recommended for any normal talk. It may occasionally need some slight extension, for example, when speakers feel it imperative to give precise figures or scientific detail (we will come back to that under 'The technical speech' in Chapter 6) but the general rule is to simplify to the A6.

Summary

Stop worrying about what to say. Start concentrating on what the listener will hear.

Prepare in three stages.

- A4 Random ideas, quickly jotted in a period of intense concentration
- A5 Listener orientated, under four main headings with supporting material
- A6 Printed prompt of half-a-dozen words.

17

Use of this method will achieve:

1. A clear and simple message
2. An efficient means of preparation, avoiding the frustration of less systematic approaches
3. Listener orientation: in a form which is suitable for a particular audience
4. The essentials now clear in the speaker's own thoughts.

3

Speaking—structure

The speaker needs to enthuse his audience. He must put his energy into that process of enthusing. Yet it is recommended that he should have only a minimum prompt with him when he is speaking.

This may sound terrifying but in practice it is surprisingly easy, provided that the speaker develops the skills.

He must have prepared systematically.

He must have a minimum prompt in front of him.

He must have a technique to help him to structure and elaborate his message.

He must develop the knack of projecting his energy to his listeners.

This chapter is concerned with the third of those elements: developing a technique which helps the speaker to elaborate without having to think about it. He needs a drill, a procedure which he can follow: a drill which will help him to articulate a clear and simple message while concentrating his energy on the audience.

This chapter aims to suggest such a drill under the headings:

1. Key moments in a speech
2. Bridging points
3. Use of humour
4. Finishing on time

Key moments in a speech

There is a consistent pattern to the way in which an audience concentrates.

Attention is at its highest for a very short period, probably not more than a couple of sentences, right at the outset.

Concentration then sinks dramatically. It will stay at a relatively low level, disturbed only by occasional waves of increased interest until the audience senses that the end of the speech is near. Then there is a sudden resurrection of interest, with people anxious to sense that they have not missed the whole drift of the talk. The final period of high interest is even shorter than the initial period—maybe just one sentence. The speaker continues beyond that length at his peril.

The experienced speaker therefore uses a procedure—a drill which he uses repeatedly—to capitalise on this pattern of concentration.

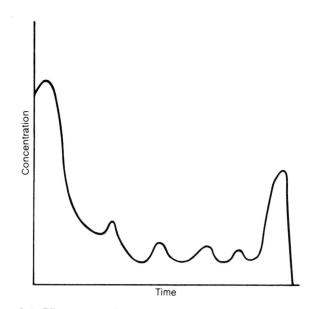

Figure 3.1 The pattern of concentration

The opening phase is of crucial importance. The speaker is giving the vital first impressions to the listeners and it is imperative that he uses his energy dramatically in this phase to establish the right chemistry with his listeners. More of this in the next chapter.

His energy should be so engaged in projecting himself to the audience that he cannot spare the effort to think about the precise words which he is using. He must have a drill to help him find those words automatically.

These opening moments of high concentration give him a chance to stimulate the expectations of his audience. He can prepare their minds for what is to come. He can feed them a sense of the direction in which he and they will shortly be travelling together.

The method at this opening stage is simple. Use the critical half-minute of high concentration to state the title and to give signposts (the headlines). The speaker has those headlines visible, abbreviated so that he can see them at a glance, on the prompt in front of him, the A6 postcard.

DO NOT:

> be afraid to repeat a title which may already have been given by the chairman;
>
> try to set the context before stating the title;
>
> try to justify your subject matter (... the subject is important because ...)

DO:

> state clearly the title;
>
> state clearly the headlines;
>
> hold back any justification or qualification until the title and signposts have been made clear. For example, 'Good afternoon, gentlemen. Our topic this afternoon is concrete bridges and our concern is with design aspects. I shall treat the subject under four main headings:
> 1. Types of concrete bridge
> 2. Their respective costs
> 3. Load factors
> 4. Current developments
> Starting then on the first of these issues: types of concrete bridge ... '

Such a simple opening is both in a pattern which any speaker can use repeatedly and also serves to alert and sharpen the attention of the listener in the opening short phase of high concentration.

The second period of predictably high interest is the final thirty seconds or so.

The listener's strongest recollection is of that which he has heard during the last thirty seconds. It is a gross waste to use that opportunity simply to fill in some detail of the speaker's final point.

The opportunity is to remind the listeners of the whole sweep of the talk. To leave them with an overall view. The method is simple. Summarise.

As a regular repetitive drill, feed back to the audience the essence of the talk, using the same headlines as were featured in the opening and as have been elaborated in the main body of the talk.

This use of the high concentration periods conforms to the established dictum of the old Army Sergeant-Major ...

'Tell them what you are going to tell them.

Then tell them.

Then tell them what you have told them.'

Bridging points

During the main part of the talk concentration will be lower than during the high spots at the beginning and at the end.

The successful speaker is, however, concerned constantly to refresh the interest and concentration of his audience.

While most of this refreshment is taking place at the personal level, to be discussed in Chapter 4, there are elements which can be done as a drill.

A particular help is the use of bridging techniques between successive sections of the talk. Pointedly make the audience aware of such bridging points.

Remind them of what they have been expecting. In the opening couple of sentences of the talk, the listener was told the headlines of what was to come. The next headline should be repeated, pointedly, making use of known and accepted words.

'... and so we come to our second main point: the respective costs of the different sorts of concrete bridge ...'

This awakens a brief eddy of fresh concentration. The eddy can be helped to swell towards a wave if the speaker subtly uses summaries and sectional signposts.

If he ends the first part of his talk by telling his audience that he is going to end the first part of his talk—then their interest will revive briefly. He can then give them a one sentence interim summary.

This must be done carefully: avoid especially the use of the word 'finally' in any such interim summary. It is a word which is heard loud by any listener, and nothing is more irritating than the speaker who is heard to say 'finally' on several occasions and still go on ... and on ... and on ...

Having offered an interim summary the speaker should then remind his listeners of the expected section heading, as above. In addition, to capitalise on this fresh impetus to concentration he may be able to whet the appetite for the next section—he may be able to offer sectional signposts.

Such interim aids to the listener must be used pointedly. It is no use letting them float past in a monotone muddled with the content of the message. They must be pointedly segregated—a pause, a change of voice, possibly the use of some illustration or reference to the headlines displayed visibly from the outset.

There are other possibilities of helping the concentration of the audience. They come either through the personal interaction between speaker and audience, to be discussed in Chapter 4, or through the use of visual aids, to be discussed in Chapter 5, or through varying methods of presentation, to be discussed in Chapter 6.

As a drill, however, the best way to refresh audience concentration in the middle of a talk is by the use of these elements of discontinuity, that is:

The interim summary

The sub-titling—repeating the pre-stated headlines

Signposts for the new section

Use of humour

Some authorities advise starting a talk with a joke, as a regular procedure or drill. They argue that this helps the audience to relax and makes the speaker more acceptable.

The same authorities believe also that it pays to send the audience away happy, smiling, over a final joke.

My experience is that when I start a talk with a joke, it produces a profound and painful silence.

So I have failed miserably to capitalise on the opening phase of high concentration.

Equally, when I end with a joke—whether or not it manages to amuse—it does not help my audience to go away remembering the full thrust of my message.

Humour is, however, helpful when it grows readily in the relationship between speaker and audience. If the speaker smiles and sees a responsive smile from someone in the audience ... and then the smile ignites such a response from those around ... such intimate use of humour between the speaker and audience, however rare, is highly productive.

But it is of the essence of such humour that it is intimate. It is not that prepared sort of humour which is the prepared joke.

For a professional comedian, of course, the opening joke can be very powerful. If you want to be judged as a comedian, open with a joke.

If you want to put a message across, don't.

Finishing on time

The effective speaker is very time conscious.

He tries to fulfil the audience expectations about the time of his talk. They will not be too concerned if, after a well-organised and well-presented talk, he finishes a few minutes ahead of time. But they become itchy if he drags on beyond the allotted span.

In the press of starting a speech, with energy absorbed in creating good first impressions, it is all too easy to ignore the time of starting; then, much later to find oneself worried as to whether one has overrun.

The recommended procedure here is to take action to

control time immediately before getting up to start the speech.

Take off the watch. Note at the top of the A6 postcard the present time and target finishing time, e.g. '8.04–8.29'. Lay the watch beside the postcard. With the simplicity of the four points on the postcard, and the watch alongside, it is then remarkably easy to pace the talk. The subconscious seems to take account of the watch and the figures at the top of the postcard so that the finishing time is reasonably achieved.

Summary

The audience concentrates on a talk in three phases.

Opening phase—high concentration

Middle phase—relatively low

Final phase—high

To capitalise on that pattern:

Feed their expectations. Tell them what you are going to tell them.

Then elaborate.

Then help their recollection: tell them what you have told them.

Help concentration through the middle:

Interim summaries

Sub-titling

Sectional signposts

Be chary of prepared jokes.

To control timing: set watch beside time-note.

Regular use of these techniques empowers the speaker to:

1. Have a drill which he can use automatically at key moments.
2. Capitalise on the audience's key concentration periods.
3. Prepare and fulfil the audience's expectations.
4. Have his energy available to enthuse his audience.

4

The personal element

The knack of the effective speaker is to ignite interest and enthusiasm in his audience.

Let us suppose that he knows his subject, that he has so prepared that the essence of what he has to say is clear and appropriate for the audience, and that he can automatically produce the appropriate words at the beginning, at the bridging points, and at the end of his talk. He yet needs the skill to enthuse his audience.

In this chapter we discuss that skill under the headings:

1. The knack
2. The technique
3. First impressions
4. Confidence and nervousness

The knack

The knack of the effective speaker is to project his energy to the audience.

The knack is to concentrate on them. To exchange glances. To recognise their uncertainties, even to create uncertainties and then respond to them. To exchange mutual understanding and mutual sympathy. To command conviction.

The knack for him is to use his energy in setting up this pattern of non-verbal communication, so much more powerful than mere use of words.

For the audience, the consequences of vitality, conviction,

enthusiasm, coming to them is an awakening of a sympathetic response; the warmth and enthusiasm then being reflected back and multiplying between the two parties.

This is the opposite of the speaker who is seen to be worried about his message, concerned about the words he is using, gazing into the remote distance while desperately striving to sustain interest.

The positive knack is in being constantly aware of the audience, of seeking—looking for—a lively response from each member of the audience.

Once mastered, the knack is surprisingly simple. It's like the knack of riding a bike: terrifying till mastered, then simple.

The technique

There are four main elements of technique in the development of this knack. They are:

Posture

Gesture

Eye-contact

Voice

POSTURE

The position of the body conveys some sort of non-verbal message.

The slouched stance, the lop-sided drooping head and shoulders. All are symbols of sloppiness.

A sagging knee, a fidgety stance, trembling hands, furtive glances, nervousness: low credibility.

Upright, shoulders back, chest out, stomach in: arrogance.

The effective speaker needs to be seen to be confident and lively. He needs therefore to stand upright, yet relaxed. He should not necessarily stand still. If he moves about a little within a radius of a yard or so, he can add animation to a sense of confidence. But if he rambles up and down the platform, he becomes very distracting.

GESTURE

The practised speaker makes a great deal of use of different sorts of gesture.

The face is expressive. If it is seen to be smiling, happy, relaxed, the audience will be responsive. If it is seen to be frowning, taut, uncertain, then again the audience will respond with like mood.

The hands can be very powerful. For this, they must move animatedly above stomach level. As long as the hands are kept below stomach level they remain powerless to influence or enthuse.

Choice of hand movements should not be deliberate. If a speaker using his energy to enthuse his audience does not worry about his actions, then his hands move automatically in keeping with the words and mood which he is projecting.

The larger the audience, the larger the gestures which are appropriate. Addressing a group of half-a-dozen people, the hands moving around chest level are expressive.

For an audience of 100 people, the forearm or even the whole arm needs to move to have an equal influence.

Other gestures have influence: the shrug, the quizzical gaze, the lifting of the eyebrow. All convey some pattern of communication, inviting the audience to respond (non-verbally), helping to infect and share the enthusiasm between them and the speaker.

EYE-CONTACT
Direct eye-contact between speaker and audience is an essential part of the personal interaction.

It has been found that there is some duration of eye-contact which people find naturally desirable. It is a shared glance of about one second duration. If a speaker fails to have such contact, repeatedly, with each individual or section in his audience, he is failing to use a very powerful weapon. But if he exceeds the acceptable duration of about one second, it quickly becomes embarrassing to both parties. (If in doubt try it with the person sitting opposite you in the train on the way home.)

Continually the speaker should have eye-contact with his listeners. He should ensure repeated eye-contact with each and every listener. (Every section of listeners, in the case of a large audience.) If one listener is temporarily out-of-contact, note-taking for example, the speaker should be specially alert for that listener: the moment he glances up, there should be a brief eye-contact. The listener should see the speaker's signal—'I'm interested in you'.

If a speaker finds himself confronted with an audience averting their gaze, he must do something about it. Not obtrusively: banging desks and turning somersaults are irritating gimmicks. Subtly, he must attract their gaze, by pausing, or by making a deliberate stumble, or by writing or drawing on a flip-chart to reinforce his words. Once having attracted the audience's gaze, he must quickly establish and then sustain the critical eye-contact.

One way to get the gaze is to use something visual: some pre-prepared display, possibly even the headlines (signposts) for the talk; or even one word written on a paper board or blackboard. The speaker must stand immediately alongside the visual aid, so that he is seen in the same glance; and he must have his attention, energy, eyes, seeking the personal contact with each listener.

He must NOT have his own attention, interest, energy captured by gazing at the board. His energy is for his listeners.

This concentration of eye-contact serves two ends. On the one hand, it is a technique for earning audience interest.

On the other, the speaker constantly seeking such personal eye-contact *externalises* his energy. He finds himself projecting. The words, the postures, the gestures: all flow. They become as natural as in a dialogue, responding to the audience's mood and reactions.

By contrast, an introverted speaker worries about his choice of word and phrase, talks to the ceiling or the floor, constantly dredges deep within himself for fresh thoughts.

The most powerful catalyst for positive chemistry between speaker and listener is eye-contact.

VOICE

The human voice has four major variables. To hold the audience's interest and concentration, each of these variables should be modulated.

The variables can be described as four Ps:

> *Pace* The speed of delivery. *A fast passage* should be followed by a s-l-o-w p-a-s-s-a-g-e.
>
> *Pitch.* Start deep down. Modulate between high and low.

Power. Vary the power of the voice. For a particularly important passage, drop the power—it can have surprisingly more influence than raising the power.

Pause. Make use of pauses. Give the audience time to collect and to digest a statement. A pause of say four seconds, which may seem interminable to the speaker, will yet be found by an audience to be a refreshing help.

In matters of technique, then, the skilled speaker makes use of the non-verbal aspects of communication—posture, gesture, eye-contact and voice variables.

Throughout his speech, he uses this technique to transmit his energy and enthusiasm to his audience.

First impressions

First impressions are of critical importance.

We know already that the listener's pattern of concentration, whatever his starting level, will soon sink drastically. First impressions decide whether the starting point from which it sinks will be a high one or a low one.

What the speaker says during these critical opening moments should be a matter of routine: the title, and the signposts for his talk. His appearance—dress and personal presentation—should be seen to complement the image he wants to project.

His energy should, however, be concentrated on critical non-verbal reactions which his audience will have.

Here are the guidelines for this critical phase.

1. Beforehand: 'prompt' card on table, time noted, watch alongside.
2. Stand up and get rid of the chair.
3. Wait.
   Wait for the audience concentration to gather. Wait until most of them are looking at you. Then look at the one or two who have not yet given you their attention. They may be chatting to one another or rustling their papers, or something else. Whatever it is, they will distract the rest of the audience and

they will distract you, unless you gain their attention.

If you ask for it explicitly, they will react against you. But if you look at them, the remainder of the audience will look with you. It will then be the power of the audience which they feel. As they look up, a brief smile from you and—with precise timing—start to speak.

4. Spread confidence.

Your opening must be seen to be confident, must be heard to be confident. Posture, gesture, tone and choice of words must work together, particularly in the crucial opening seconds.

Never never start with an apology ('I'm sorry that Mr Jones could not present this paper today'), nor with an implication of doubt ('I hope that you will find it interesting ... ').

As previously advised, confidently open with title and signposts; and reinforce the audience's confidence by the manner in which you state them.

Confidence and nervousness

Inexperienced speakers are often highly concerned about their nervousness and their apprehension immediately before they have to give a talk. They seek guidance on how to relax at that time.

I am by no means convinced that they should.

There are indeed those who perform best when they are relaxed. For them, the advice should be to take the normal steps towards relaxation. Relax the muscles, drop the shoulders half an inch. Feel each finger relaxing, one after the other, then each hand. And so on.

But in common with many other experienced speakers, the last thing I want to do before giving a talk is to relax. I want the adrenaline flowing. I want my concentration to be at a high. I deliberately take steps to concentrate—digging my finger nails into the palms of my hands, breathing deeply. When I recognise that the moment is at hand, I sit well upright so that I can get a good lungful of air. Then I take three very deep breaths.

Thus, when I first stand up I am tensed up. I think in the last few moments before standing, of the words 'audience-contact'. Then I stand, wait for my audience to collect themselves, then start, confidently.

That confidence springs from having systematically prepared the speech, and from knowing how to structure it. And it may need a bit of acting too—don't be afraid of it! Be seen to be confident.

Summary

The knack of the good speaker is to use his energy to enthuse his audience.

Equipped with a clear view of the essentials of what he has to say, he addresses himself wholeheartedly to capturing their interest.

He makes use of non-verbal technique: confident posture, animated gestures, modulated voice.

He makes powerful use of eye-contact, both to help his listeners and to help himself to externalise.

He takes special care to create good first impressions: he pauses to collect his audience, and generates confidence with the way he makes his opening remarks.

He organises his nervousness. He does not worry about having butterflies in the stomach. He gets them flying for him.

5

Visual aids

Things which the listener can see have a powerful influence on him. They capture his eyes and his attention: they affect his concentration and his liveliness. They can be a great help or a great hindrance.

In this chapter we will be concerned with three main types of aid, and some suggestions on how to use them; boards, slides, and the overhead projector.

But first, a few general rules.

General rules

Visual aids are very powerful. They can dramatically increase the interest and attention which an audience is giving.

They can therefore be very powerful servants. They can also be powerful opponents. An aid which is badly chosen or badly used can distract the audience.

Aids can also distract the speaker himself. They can magnetise his interest, attention, gaze, so taking his energy away from the essential audience contact which he should be establishing.

To be effective, a visual aid must be simple. It must help the viewer quickly to perceive and digest a message. If the aid is cluttered with too much detail, it is a distraction for the viewer.

It should be bold. Strong lines, large letters, so that it can easily be seen at whatever distance the rearmost viewer will be sitting.

Avoid distractions. Viewers invariably concentrate on

them. If you want listeners to see a picture of a ship on the screen, you might well find that the photographic balance would be better if there was a yacht somewhere in the background. In that case, they would see the yacht and ignore the ship. To get attention to the ship you might put a young lady, possibly in a bikini, on the deck. They would then concentrate on the young lady!

Visual aids may be pre-prepared, or may be developed while the speaker is talking—he may be writing or drawing to illustrate his point.

The pre-prepared visual is seen by an audience to be brought in from outside the meeting place. It carries an element of prestige and authority from the outside world. It can therefore enhance the speaker's own perceived authority. It can help to reinforce his status as an expert on his topic. For this, of course, it must be seen to be neat, tidy, elegant— and, at the same time, bold and simple.

If, on the other hand, a speaker wants to establish an intimate rapport with an audience, then they will be more ready to accept something which is actually created in their presence.

Talking to an audience, to impress with one's expertise and authority, use pre-prepared material. Working with a group, create it as you go along.

When using pre-prepared aids, display them only at the appropriate moment during a talk. If they are visible before or after that time, they will powerfully distract attention.

Always check the visibility of such aids before the meeting takes place. Look to see that the lighting is suitable, that the writing is big enough and bold enough, that the marker pens will actually work when needed.

Most particularly, concentrate on audience-contact. Avoid being magnetised by any aid and work hard at the eye-contact and other patterns of behaviour which help you to externalise your energy and to involve your listeners.

Boards

There are three types of board which a speaker can use to write on. Blackboards, white boards and paper boards.

The traditional blackboard has just one advantage. It is

cheap. However, the chalk is messy to use, the surface is liable to reflect light unhappily and, often enough, it is very difficult to read.

White boards are preferable. They avoid mess. Provided that care is taken to use only washable inks, the board can easily be cleaned. Different colours of felt-pen can be used. Reflection is still a problem, but visibility tends to be better than with the blackboard.

The paper board or flip-chart is typically of size some 45 cm × 65 cm with each sheet as it is used being turned over the top of the board. It has all the white board's advantages over the blackboard, and other advantages too:

1. It is easy to prepare some sheets in advance.

2. Sheets may be shown or hidden at will.

3. The speaker can turn back to remind listeners of a sheet previously displayed.

4. A sheet can be removed and durably displayed, more or less prominently on a wall. (Use masking tape or Blu-tack to hold a sheet in place with little or no risk of damage to the decoration.)

The paper board is accordingly the most desirable among the three types of board.

Any of the boards can be used against a wall or free-standing on an easel. The latter—the free standing—is much preferable. It enables the speaker to write side-on to the audience, whereas the wall mounted board forces him to write back-to-audience. The wall mounted therefore interferes very heavily both with voice projection and with eye-contact.

When making concurrent use of a free-standing board (writing or drawing on it during the talk), the best position for the right-handed speaker to stand is audience-left of the board. He can then write so that he obscures very little of what he is in the process of writing. It is much more difficult if he starts writing from a position at the opposite side of the board.

Displays can be prepared for boards—for paper boards, for flannel boards or for magnetic boards. When using these devices bear in mind the need to be simple and clear in what is being stated; to be simple and bold in constructing the aid.

Keep it obscured until it is needed. When displayed, give a brief moment for an overall impression, then describe simply what the display is meant to convey, drawing the listener's attention to the key items and specific units of measurement being used. Do not expect him in a flash to absorb the whole picture without adequate explanation. Give him time and help.

Slides

Effective use of slides can be a great help to listeners. They can sustain interest and attention as well as convey information.

They can also be ruinous. They get put in upside down; they can be in the wrong order; the speaker has a private device (possibly the dreadful tap tap of a foot) to signal to a projectionist, and when it is all over, the lights suddenly come on and blind everybody.

To make more effective use of the possibilities of the slide projector:

Keep the message simple on each slide.

Code the slides to show how to insert them.

Index for correct sequence.

Dim and later brighten the lighting gradually.

If possible, let the speaker sit by the projector in the middle of the audience. Then—

Adjust voice from the authoritative standing, to the friendly intimate, position of being a member of the audience.

For a pointer, use an arrow-headed torch.

The overhead projector

The overhead projector is extremely popular—excessively so, in my view. It enables a speaker to face towards his audience and to project prepared transparencies on to a screen or wall behind his head. Alternatively, to write on a continuous transparent roll with the flexibility to wind forward to a clear sheet, or backwards to an earlier display.

It is a powerful and very versatile tool. It gives great scope for ingenious development of displays and can even be used for animated projections.

Powerful as it is, it is usually a powerful competitor. It has a fan which may make a distracting noise; and, more important, the image is appearing to the listeners at a point away from the speaker. The chemistry of the interaction between speaker and listener is jeopardised. At the same time the projector is a physical barrier between speaker and listener, partially obscuring him and certainly acting as a psychological barrier. Because of the barrier, the speaker tends to concentrate energy on the equipment. There are even authorities who suggest that he should point out the details with a needle or pointer on the transparency, but this emphasises the distraction of the image remote from the speaker.

How then should the speaker make use of this powerful tool?

In general, avoid using the projector for concurrent writing and speaking. The overhead projector is too tough a competitor in this situation. Much better use a free-standing paper board or white board.

When using a pre-prepared transparency, stand by the wall or screen so that you are seen to be integrated with the display. To point out details, use a pointer to a position on the screen and not a pencil or needle to the transparency.

It is possible to use a great deal of ingenuity in the preparation of displays. Shortly there follows a series of hints about devices which can become tools for such ingenuity. There is, however, a great danger in this; the danger that the speaker's energy will become captured by the device. He may find that he is so concentrating on the projector, the intriguing transparencies, and the absorbing image on the screen, that he loses contact with his listeners.

Provided that the talk is prepared for the interest of the listener, it is possible to use the versatility. It can be used for:

Plain writing

With the photocopier, for typed documents or drawings

Using coloured inks, for colour differentiations

Parts of a display can be masked, while other parts are being elaborated on

Successive detail or development can be shown by build-up of a series of superimposed transparencies. If wishing to do so, hinge them together with adhesive tape.

Despite the general advice against using for concurrent writing, it is nevertheless possible to position transparencies under the continuous strip of clean transparency and thus to write over a master.

Cut-outs can be made, displayed, moved, animated.

For dramatic impact, prepare a carbon-coated transparency and chalk a figure over the carbon. During the talk, drag a needle across the chalk outline. The listeners' perception is of a light beginning to shine through the lines in the middle of blackness.

Small objects can be projected in silhouette. Even live objects, insects for example, can be projected if inside a transparent case.

A card or sheet of paper with a disc or rectangle cut out can be used to focus attention on one area within a display.

Different coloured cellophane can be used.

Animation can be generated by over-lays, bars pushed up or down, discs rotated, cut-outs manipulated.

These give great scope for creative imagination, but beware of the temptation to let the tool matter more than the listeners.

Summary

Visual aids are powerful in communication. They can be powerful aids or powerful adversaries.

Keep them simple and bold. Concentrate in preparation on the listeners' needs and on using the tool only as an aid to fulfilling those listeners' needs.

Concentrate during presentation on audience-contact and not on the tools.

Check your position so that you are seen to be in harmony with the aid and not in opposition to it.

Use free-standing boards rather than wall boards; paper or white boards rather than blackboard.

Keep slides simple and make sure that they are adequately explained.

Beware of the overhead projector. Use principally for pre-prepared displays. Make creative use of the aid's great potential, but only as a servant—don't let it become the master.

6

The major speech

Previous chapters have suggested a technique for giving a talk. It has been assumed so far that the topic is one of general interest, to be dealt with in a length of time for which listeners can readily sustain concentration: possibly about 20 minutes.

Engineers can be called upon for different sorts of speech. They may be required to make longer speeches, or to make highly technical presentations. They can be asked for the sort of speech to which eminent colleagues will listen and on which a judgement will be formed of the speaker, his technical competence, and the competence of the organisation which he represents.

The speaker then is likely to come up against the hazard of the large auditorium with its battery of microphones and the hazard of handling questions and discussion.

In this chapter we move on to techniques for handling these situations.

At the outset let us distinguish between four types of speech:

1. A talk. A topic of general interest, without heavy technical detail, with duration of possibly 15–20 minutes.

2. A major speech, for example, a conference speech, possibly to fill a one-hour slot.

3. A technical presentation in which it is imperative to communicate technical details.

4. A published statement.

This chapter will deal with the distinctive features of technique for the last three of those situations—major speech, technical presentation, published statement.

The sequence of the chapter will be:

1. The design of a long speech
2. The technical speech
3. The large auditorium
4. Use of microphone
5. Questions and discussion

The engineer also attends conferences in the other role: that of the ordinary attendant, who may ask questions or contribute to discussion of someone else's paper. Behaviour in that role will be discussed in the context of 'Bigger meetings', Chapter 21.

The long speech

This section is concerned with the issues:

How long should a long speech be?

How should it be designed?

How should it be prepared?

First, the length of a speech.

There is a span of time over which people can listen with good concentration.

The length of time depends on a lot of factors. The skill of the speaker, the time of day (longer in the morning than afternoon or after dinner), the interest of the audience in the subject matter, the chemistry which develops between the speaker and the audience.

Despite these variables, however, there is some length beyond which listeners cannot be expected to give their full attention. This time is of the order of 15–20 minutes.

Gifted speakers, working with interesting material in good surroundings, may be able to sustain interest for as much as 35 or even 40 minutes. Not longer. (How many students sustain concentration throughout a 50 minute lecture at university?)

The engineer thus is setting himself a stiff target if he aims to hold listeners' concentration over a span as long as 15 to 20 minutes.

So how should he respond to requests for an hour-long talk, to be followed by a question session, at a conference?

Ideally, he should negotiate with conference organisers, so that his contribution should not be expected to be more than the listeners' concentration time. But the nature of conference planning seems to be such that speakers are under pressure to perform for an hour, followed by a half-hour discussion.

Given this demand, the speaker should take the time he has to fill, and break it into shorter periods. In each period he should choose a radically different technique for his presentation; thus regularly revitalising listener attention and concentration.

For example, an ideal breakdown of a one-hour period would be into four sections, each expected to take between 10 and 20 minutes.

Section 1 might take the form of a formal speech, the speaker standing at the front of his audience, facing them. The atmosphere would be one of authority emanating from the platform.

Section 2. Change the atmosphere. Use visual aids, for example, a slide projector. Move into the body of the hall to use the projector and talk in a more intimate way from the audience. Work together with them, using the arrow torch to help, concentrating together on the screen.

Section 3. Involve the audience. Invite them to consider the questions they would like to raise at this stage. Or ask for other comments and experience. Or break the audience into small groups to discuss and subsequently report back on implications of what they had so far heard.

Section 4. Some other format, continuing to create new attention. For example, gather three or four prominent questioners from the audience; invite them to join the speaker in a semi-circle at the front of the hall to discuss the topic with one another, in public.

Such discussion not only provides a new focus for concentration, it also excites enthusiasm from the general audience, who feel a fresh impulse to debate with and question their peers previously in the audience. They do not like being left out.

Finally, summarise the essence from the paper and from the subsequent discussions.

For preparation methods for such a major conference session, follow the methods suggested in Chapter 2.

The A4 stage of random jottings

The A5 stage: listener orientated, organising under a small number of main headings

The A6 stage: reducing to a minimum prompt.

Most speakers need to use this method repeatedly when preparing a major session.

First, as a procedure for planning the full presentation in some four separate sections. The topics for these four sections need to be chosen, bearing in mind both the interests of the listeners, and the need to change the form of presentation three or four times, to revitalise attention.

Second, repeated use of the A4/A5/A6 technique, once for each section to be covered.

The end product of this process will be five A6 cards, the top one bearing the title and the four sections to be covered; backed by one card for each section, each card bearing the four or so headlines for that section.

Having got the plan thus clear, the speaker will need to be sure that he has suitable aids—especially visual aids—for his presentation. He needs to take the time and trouble to ensure that he has the appropriate models, drawings, or slides. Accurate, simple, visible, in the right order.

Summarising key issues for a major contribution to a conference:

Listeners cannot be expected to concentrate for more than 15–20 minutes at a time without special help.

Plan for positive variety. Revitalise concentration by changing presentation techniques three or four times during the session.

Use the recommended A4/A5/A6 approach repeatedly to prepare a session.

Have suitable aids prepared to illustrate.

The technical speech

Listeners cannot easily absorb highly technical information. They cannot easily absorb a lot of facts and figures. The engineer who is determined to give all the evidence in a talk, is thus doomed to failure. He will not be fully understood.

The content of such a speech should deal with the definition of the problem then with the statement of methods used and problems encountered. Do not try to feed all the results. Restrict yourself to a summary. Thereafter, discuss the significance and draw conclusions.

To the extent that some precise information must be conveyed, the speaker needs to have it written down. He can if necessary write it on the back of his A6 card. But the extent of such precise detail should generally be restricted. As a working guideline: restrict it to one line of figures.

When he believes it is of key importance that such a line should be communicated, he should emphasise both by making it visible (writing it on a board) and by articulating it at least twice. Give the listener plenty of chance to hear, to see, to understand and to digest.

If a speaker feels it imperative to present more detailed specific results, he should of course have them pre-prepared on a chart or slide. He can display these slides, he can read from the information on them, and he can state the conclusions that he draws. He should not, however, expect everyone, even within a technical audience, to follow all he is saying.

He can succeed in:

1. Pointing out exceptional figures or orders of magnitude
2. Displaying the scope and the character of the work he has done
3. Claiming credibility
4. Pointing out key conclusions

He cannot succeed in:

47

1. Getting full understanding of his results
2. Getting what is understood, digested then and there
3. Enabling listeners to draw their own conclusions from the results

General guidance, from a communication point of view, is therefore severely to limit the quantity of technical detail in a talk.

There are, however, occasions when the engineer has to present a wealth of technical detail to colleagues, both for their information and as a stimulus to criticism and further development. In such cases it is essential that the message should be put in writing and should be pre-circulated. Conference members will then have had a chance to read, to re-read, to work through some implications, and to digest the arguments before the conference.

Conference members will have had a chance to read, but many of them will not have read the pre-circulated paper. The speaker now has a difficult situation to resolve.

On the one hand he cannot hope then and there to convey the full detail which he has written, which some of his listeners will have studied carefully. On the other hand, he must present sufficient for the non-readers to enable them to follow the gist of the paper and possibly even to participate in discussion.

The need is for some synopsis, giving an outline without going into detail. The headings might be, for example:

Definition of problem

Scope and methods of work done

Salient results—restrict to three or four key factors

Implications

The general guidance given in this book is to prepare systematically and to let the words which the speaker uses develop responsively to the mood of the audience. This technique frees the speaker's energy for very positive audience contact.

It is not always possible. If a speech is to be published then it may be that it will attract critical examination. In that case, every word needs to be weighed, every statement checked.

In most learned societies it is possible to pre-issue a statement or paper for publication. When the time comes for him to speak, the author can ask if the paper can be taken as read. Any report should then be of the writer's statement, and the speaker can have more freedom in the words which he uses in giving his oral synopsis of the paper to his audience.

If circumstances force him to read a written paper, then the need for audience contact is by no means lost. He must use his voice carefully, pausing and varying his pace, power and pitch. He must have the papers at a sufficient height so that his head is not bent down to read, and so that the voice can project. He should be constantly looking up, searching for eye-contact and he must be positive in his use of gestures.

For the most important audiences, he may even go to the extent of using a teleprompter or autocue—the device used by television newsreaders, so that viewers believe the reader is looking straight at them through the camera. In fact there is an invisible screen between reader and camera, with the news being continuously wound on to the screen.

The speaker should show energy and animation. The obstacle is that the written word can mesmerise his attention and energy. The consequence then is a dreary monotone, a soporific presentation.

Avoid it. Concentrate instead on animation, liveliness, audience-contact.

To summarise: listeners cannot absorb much technical detail in a speech. Keep it simple. Keep it short.

If that cannot be done, use writing rather than speech to communicate.

Even when forced to read, still strive for animation and audience-contact.

The large auditorium

The ground rules for speaking in a large auditorium are the same as for speaking to a smaller audience.

Prepare systematically.

Orientate to the listener.

At the beginning, create expectations.

Through the middle, fulfil them.

At the end, summarise.

Be highly aware of the audience.

Use energy to enthuse.

There are, however, differences of detail in the presentation, particularly those aspects which we earlier called the personal element.

Eye-contact is again of great importance. It is not possible of course to have individual eye-contact with each member of an audience of 400. But in practice, with such an audience, even the most gifted speaker finds a great many apparently poor listeners: those who do not offer eye-contact but sit with eyes closed, or glazed, or focused on the window. Some sit slumped and inanimate.

The speaker should then ensure that he does hold contact from time to time with each section of the audience. Upstairs and downstairs. Left, right and centre.

If he is concentrating on this audience-contact, he will soon recognise one or two members in each section of the audience who are lively listeners. He can build the chemistry of delivery through the contact with those individuals. At the same time, he will be seen to address himself not to the individual but more generally to the section in which that individual is sitting.

Gesture in the large auditorium must be more expansive. Where the scale of gesture with a small audience should be 'hands and fingers', in the large auditorium it must grow in scale to become 'wrists and hands' or even (for instance in the open air) 'arms and hands'.

Voice projection must be adjusted, both in pitch and in power, to meet the larger scale, or to meet with that further hazard of speakers, the microphone.

Use of microphone

The microphone is indeed a hazard. It is likely to screech, to whine and then to break down.

Even when the microphone is working efficiently, the speaker is seen to be speaking somewhere and the sound is heard to be coming from elsewhere. This is hostile to the speaker's audience-contact, hostile to his concern to enthuse the audience with his energy.

In general, therefore, I do my utmost to avoid using a microphone.

When obliged to use one, whether because of the gross size of auditorium or because of the insistence of conference organisers, the speaker should take three practical steps.

1. Pre-test. Arrive in good time for the speech. Get a friend to come along and sit where he can criticise. Test the microphone for position and for volume.

2. Do not adopt the pop singer's technique—mike to mouth. It distorts and makes the speaker unintelligible. Keep the microphone 6 to 9 inches from the mouth.

3. Direct the microphone at the chin or throat and not directly at the mouth.

Those are aids to the technique of using a microphone; but the basic advice is to avoid it where possible.

Questions and discussion

Interest and intensive discussion are an accolade for any speaker.

The potential for such interest and intensity depends partly on the process used to control the discussion. It is usually controlled not by the speaker but by a chairman who can make or mar the occasion. The subject of conference chairmanship is therefore considered as part of Chapter 21. If the speaker should be in the fortunate position of himself being able to control the discussion, the advice in that section will be relevant.

Engineers are often concerned as to how they should answer questions which arise at the end of a speech. This depends in part on the sort of question.

There is the helpful question. The one which is put clearly, which is relevant, and to which the answer is reasonably straightforward. No problem.

There is the mumbled question. The speaker is not sure what has been asked.

He can be absolutely sure that the rest of the audience has understood even less. Help them. State clearly, for their benefit, the question which you think should have been asked. Then answer that question. If too uncertain of the mumble to try that procedure, then ask either the chairman or the questioner to repeat the question.

There is the 'murky' question. The one which is obscure because the questioner has used phraseology which the speaker finds difficult to understand.

If he finds it difficult, the rest of the audience will also find it difficult. What is more, if the question is not clearly stated, it is likely that the questioner is not clear in his own thinking. Both the questioner and the rest of the audience, and ultimately the speaker himself, will be helped by a re-phrasing. ' Thank you—may I just make sure that I have the question right in my own words. As I understand it, the question is … ' Then quickly, start to answer that question—don't give him the chance to comment before you do so, or he will inevitably add to the obscurity he has already created.

There is the peripheral question. The one in which a questioner raises a side issue or a remote issue, where the speaker would be on doubtful ground in replying.

Use honesty and integrity. 'The important parts of the evidence of our work for that are … However, it is not a subject on which I personally have great experience and I cannot be certain how that would react in the questioner's circumstances. I wonder if anybody else present has the experience or if indeed the questioner himself can help us?'

And then there is the clever question: the one which has been designed to set a trap or to be critical.

The speaker may need to repeat the question to gain thinking time for himself, or possibly to re-phrase it into a form which he can more readily answer.

If trying to answer, however, there are three types of approach, among which he must quickly choose:

1. Repeat a saying from the main body of the talk, patiently and with conviction. Then ask the clever

questioner if he would kindly test and practise and write to you with the results.

2. Feed the question back to the questioner. 'As I understand it, the question is ... It is a most interesting question. Before I comment from the background of our work, I wonder if the questioner himself has any experience to offer us on this matter?'

3. Appeal for help. 'I am not sure I can answer that. Can anybody else offer advice to our colleague?'

For the engineer, not normally a political animal, honesty, integrity and sincerity are the watchwords for his answers to questions. Professional colleagues accept and appreciate answers which have a reasonable professional degree of reservation, rather than over-confident answers to difficult issues.

Summary

Long conference presentations should be broken into sections.

Different forms of presentation should be used for each section.

The audience should be involved as far as possible.

The preparation processes should lead to a series of cards and visual aids.

The quantity of technical information in a speech must be severely restricted.

Pre-circulated papers should not be fully read, they should be summarised.

Technique in the large auditorium should be more expansive than in the small.

Microphones should be avoided or used with great care.

Honesty and integrity are the cornerstones for answering questions in most professional situations.

7

Effective listening

This book has so far been concerned with the skill of sending an oral message; but, however good the transmitter, it is effective only in partnership with a good receiver. This chapter is therefore concerned with the converse of effective speaking—effective listening.

The chapter is in four sections:

1. Hindrances
2. Positive listening
3. Interaction between listener and speaker
4. Listeners' questions

Hindrances

It takes energy to concentrate on hearing what is being said, to concentrate on understanding what has been heard, and to make an objective evaluation of what has been understood.

The energy is not deployed by incompetent listeners. They fail in a number of ways.

First, they may drift. They follow private side-tracks, with their attention drifting from what the speaker is saying.

Second, they may counter, constantly trying to find counter-arguments to whatever a speaker may be saying. This applies equally to the thought processes of people listening to a set talk and to people engaged in a dialogue.

Third, they compete. In a dialogue they seek to impose their own bigger and better anecdotes. To impose their own experience and own values with low regard for the other party.

Fourth, they filter. They exclude from their understanding those parts of the message which do not readily fit with their own frame of reference.

Fifth, they distort. Even when they receive a message they interpret it in ways which belie the speaker's intentions.

Finally, they react. They let personal feelings about speaker or subject override the significance of the message which is being sent.

There are then a series of hindrances which may prevent listeners effectively receiving a message. What can a listener do to be more effective?

Positive listening

The first key to effective listening is the art of concentration. Concentration is partially a matter of attitude. If a listener positively wishes to concentrate on receiving that message which a speaker is trying to send, then his chances of success are high.

It may need determination. Some speakers are difficult to follow, either because of voice problems, or because of the form in which they send a message. There is then particular need for the determination of a listener to concentrate on what is being said.

Concentration is helped by alertness. Mental alertness is helped by physical alertness—not simply physical fitness, but also positioning of the body, the limbs and the head. Some people also find it helpful to their concentration if they hold the head slightly to one side.

Personally, my basic competence as a listener is normally poor. My time-span of full concentration is very low. My technique for overcoming this is intensive note-taking, and my method is to try to recreate the structure of the speaker's talk: to try to capture the critical headings and sub-headings from which he is talking.

It may be of course that he is using a technique for preparation and presentation which is radically different from the A4/A5 approach outlined in this book. That does not matter.

It in no way invalidates my attempt to capture what he is trying to say in the form which I find most intelligible.

Interaction between listener and speaker

A positive chemistry between listener and speaker can enhance a listener's effectiveness. Much stress has already been laid in earlier chapters, on the need for the speaker to use his energy to create that chemistry.

The listener has an equal responsibility if the interaction between the two is to become good. The normal battery of non-verbal communications is relevant, and eye-contact is particularly important.

Some people claim that they listen most effectively when their eyes are closed. They claim that this is an aid to their concentration.

I am never sure whether to believe them. But even if one takes the claim at face value, the net effect remains bad. The listener with his eyes closed is not seen to be concentrating. He is not offering the same evidence of synergistic listening as the more obviously attentive listener.

Better practice is to look at the speaker, seeking for positive eye-contact with him, and to go beyond that to gestures which support him. Facial gestures can be particularly helpful, the raised eyebrow, the critical smile, the affirmative nod, the negative shake, the puzzled frown.

The practised speaker, recognising one or more listeners in his audience who are giving him such signals, quickly takes note. He adjusts the pace and pattern of his speech to the needs shown by these listeners, and the interaction can then become a powerful booster for both parties.

Posture too is important. An upright posture helps a listener's concentration. At the same time it is seen by the speaker to be a positive feature amongst his listeners. Contrast the impact made by a less competent listener who pushes his chair backwards and slouches.

Note-taking has been recommended as an aid to the listener. It also helps the speaker. It gives him confidence when he sees that listeners are sufficiently interested to take notes; the patterns of eye-contact when the note-taker looks up can be very positive; and the speaker's timing is aided—he can see when a note-taker is writing hard and can then make effective use of pauses. Such pauses, instigated by the needs of a listener, can often be protracted and contribute powerfully to the speaker's authority.

In a dialogue, heard-behaviour is also important. There is nothing more boring than a dialogue in which one party is constantly using his own frame of reference, talking about his own experience, his own anecdotes, his own fancies— imposing his own frame of reference on everything which the other says.

The sympathetic listener concentrates on the speaker's line of interest. He tries to understand the speaker's framework. He uses nouns and verbs in the same constructions and in the same tenses. He uses the same terminology. His questioning technique also is sympathetic.

Listeners' questions

In small groups in the normal process of discussion, or in large groups after the presentation of major papers, the listener has the chance to pose questions.

The way in which he puts his questions is important. It is possible to phrase a question so that the speaker's thinking is compressed into the questioner's frame of reference. In general this follows when a question starts with a verb. Almost automatically it pushes the speaker into a yes/no sort of response.

Consider, for example, the likely responses to questions starting with the following:

> 'Do you think that ... ?'
>
> 'Have you tried ... ?'
>
> 'Will there be ... ?'

On the other hand, questions beginning with any of the six good servants of Kipling, demand that the speaker does his own thinking. That is, questions beginning with any of the six words:

> What
>
> Who

How

Where

When

Why

Compare:

'Will the results hold in all temperatures?'

with:

'What influence will temperature have on results?'

'Has the speaker considered the influence of daylight?'

with:

'What influence would daylight have?'

'Can it be done in a week?'

with:

'How long will it take?'

There is further a difference between questions for clarification and questions for justification.

Questions based on the word 'What' are usually questions intended for clarification.

'Can you tell us what happened with ... ?'

'What would be likely to happen if ... ?'

'What are the influences on this of ... ?'

'Could the speaker help us to understand more of what is meant by ... ?'

These are all positive questions, enhancing the understanding between speaker and listener.

A quite different relationship is established by challenging questions—ones demanding that the speaker should justify. Such questions generally start with the word 'Why' or 'How'.

'Why is it said that ... ?'

'Why would X result in Y?'

'How would the speaker put that into practice?'

'How were those observations made?'

Questions for clarification have a tendency to unify the parties. Questions for justification, a tendency to set the parties on opposite sides.

In addition to their use of questions, listeners have a possibility of checking the accuracy of communication. They can do so by using the technique of *feed-back*.

The speaker has sent a message; they have received a message. It is never certain that they have received an identical message. Indeed this is very rare.

When in doubt, do not just ask the speaker to repeat. He does not know what the blockage is between what he has already tried to say and what the listener has understood or failed to understand.

Instead, feed back to him what you have understood, in your own words. The phrase for this is: 'Please may I check. What I have heard is ... ' followed by a brief summary of the headlines that one has received.

The listener thus gives the speaker a chance both to check the message and to confirm or modify.

Summary

Effective listening depends on:

1. Positive attitudes: Seeking to understand the speaker's frame of reference, not to impose one's own

2. Interaction: Using the panorama of non-verbal technique to build synergy with the speaker

3. Concentration: Often helped by note-taking

4. Questioning technique: Clarification different from justification, Open-ended questions (What, Who, How, etc.), Use of feed-back.

Part 2

Effective writing

Introduction to Part 2

In Part 2 of this book we shall be concerned with the subject of effective writing.

The opening chapter of this part is on the subject of preparation. Guidance here (Chapter 8) follows general principles discussed earlier in the book.

Writing style is the mechanism through which the writer projects character to the reader. The way he writes influences the way the reader is motivated to understand and to accept a message. Writing style is the topic of Chapter 9.

The structure of letters is the topic for Chapter 10 and in Chapter 11 we go on to look at structures for a number of standard reports and their presentation.

Engineers are often involved in the production of major reports; thick volumes, needing weeks or even months of preparation and drafting, carrying heavy influence with clients, and often being the only visible product of much client expenditure. The production of such major reports is the topic of Chapter 12.

This part of the book closes (Chapter 13) with brief guidance on the converse of writing skills: reading skills.

8

Writing: preparation and dictation

The following guidelines for preparation for writing are straightforward and follow the pattern recommended in Chapter 2 for preparation of speeches.

At the outset, ensure that the homework has been done. The facts and figures collected; the previous files found and read; the questions or statements in correspondence studied.

Thereafter, organise the thought process in two stages:

An A4 stage to get random thoughts on to paper

An A5 stage to bring order

Chosen to be what the reader will want to read

The essentials highlighted in a few bold headlines

Each headline supported by material of interest to the reader

There is one major difference in preparation. The written word can be re-read whereas the spoken word cannot normally be re-heard. This means that there can be some relaxation of the rule of organising under four main headings. (Each with its four sub-headings.)

The reader can re-read and therefore may be able to absorb a greater variety of headings. This is particularly true if a document is following a standard format, such as a technical report which typically may follow some 10 standard headings. (As in Chapter 11 below.)

But where the writer must create a novel structure to tackle a broad subject, the four-heading discipline is again important. Partly in the reader's interests, but equally for the writer himself. Unless he so sharpens his own thinking he is likely to have difficulty in drafting, and his product is likely to be muddled and murky.

Or staccato. Many engineers' letters, for example, confirm discussions which have taken place. If there have been a dozen points under consideration, then the resultant letter may consist of a dozen one-sentence paragraphs. It reads like a dozen staccato shots.

It becomes more interesting and more effective if the separate points are grouped under sub-headings.

If the writer wants to dictate, he should do so from the A5. His preparation should have helped him so to clear his thinking that dictation is straightforward. The message and the words to convey it should flow simply from the prompts on the A5 in front of him.

Relationships with secretaries are important. I meet many engineers who tell me that they dictate into machines and that the machines are then transcribed by people of low competence and low motivation. And in the same organisation I meet many typists who tell me that they work only for unfeeling engineers, who never have the interest or courtesy to discuss the presentation, the layout, the styling of the written word. Working in combination, engineer and secretary can produce a product much more interesting—in style, in layout, in appearance—and much more rewarding for them both than the results of personal warfare.

Guidance about preparation for writing is thus straightforward.

Use an A4/A5 approach

Dictate from A5

Discuss presentation with secretary

The guidance may be straightforward. It is nevertheless critical. The writer who has prepared his own thinking, and has structured it appropriately for his readership, has an excellent foundation on which to build.

9

Writing style

The engineer's writing style depends on the way in which he makes use of his basic education in English language and grammar. It is not the function of this book to provide such a basic education; but it is the function to suggest how best to make use of the education.

The sequence of the chapter will be to discuss:

First, some key variables which characterise writing style

Second, suggestions of acceptable standards

Third, use of layout

Fourth, variations for different readers

Finally, economy of words

Key variables

Four key variables which characterise writing style are:

Length of statement

Choice of words

Use of verbs

Choice of pronoun

LENGTH OF STATEMENT

First, there is the issue of length: length of sentence, length of paragraph, length of section.

Long passages are harder to understand than short ones. The longer a sentence, in general, the more difficult it is for a reader to understand that sentence. This statement must of course be surrounded with caution: there are subtleties in the choice and structuring of the words, and about the sort of relationship between writer and reader which require care. But in general the longer the sentence the more difficult it is to understand.

The majority of the population can understand sentences averaging between 16 and 19 words. The writer who uses longer sentences increasingly runs the risk of not being understood. If he uses sentences averaging more than say 28 words in length, less than 20% of the population would be able readily to follow his writing.

Consider, for example, the following extract:

'On installation of the new premises adjacent to the wharf as required by the designs submitted on 10 January and as revised at the request of the supervising authority by drawing of 28 March, the approach road from the south west side will be faced not only with the traffic volume of 943 units originally estimated in the May figures, but with the further flow which will arise because the port authorities are by then expected to be requiring discharge from the western approach as well as the new volume going direct into ... '

Compare with:

'Traffic flow: On latest plans (28 March) there will be heavy traffic flow following completion of the warehouse. This will create problems on the south west approach road. Predicted flows at that period will be a total of

(a) 943 units as originally predicted;
(b) an additional 437, due to new routeing by the harbour masters;
(c) an extra 41 units for the warehouse itself.'

In these examples, the convoluted and indigestible mass in the first long sentence is reduced to simplicity by breaking it up into a series of sentences.

The opposite of excessive length is also deplorable. A series of very short sentences reads like a series of shots at a rifle range. There's nothing attractive in them. For example:

'The latest plans are dated 28 March.

They imply that there will be heavy traffic flow.

This will follow completion of the warehouse.

There will be special problems on the south west approach.'

This is excessive reduction in sentence-length. The need is for sentences lying between the extremes of exhausting length and cursory inelegance.

Paragraph length is equally important. A guideline is to consider a sentence as conveying one thought, and a paragraph one idea. If an idea consists of more than three or four thoughts, then the writer has not sufficiently broken down his ideas. He is trying to convey ideas which are more complex than a reader will easily understand.

There is a similar need to break down sections of writing into digestible passages. The ideal is to aim for the main part of a section to consist of four to six paragraphs. That main part should be preceded and followed by one-sentence paragraphs acting as bridges between different sections. More on 'bridging' in the subsequent chapter on report writing.

Length of statement is therefore one key variable in writing style: length of sentence, of paragraph, of section.

CHOICE OF WORDS

Familiar words convey sharp messages. Unfamiliar words convey hazy messages.

There is one range of words which everybody uses daily. They are words familiar both to specialist and to layman. There is a further range of words which engineers can use with one another—words which they do use with one another every day—but which are not familiar to the layman. These are words in the engineer's jargon.

There are further words which both engineers and laymen can recognise but which are not part of their everyday language. These words, as they get further and further from daily use, become words which do not project a sharp image. They also intercept the steady digestion of a message. The reader sees a word whose ambiguity affronts his appraisal, analysis and understanding. And so, such words impede his sympathetic reading.

69

There are some 3000 words which are in everybody's day-to-day vocabulary. The educated professional has a latent vocabulary which goes beyond that 3000 to some 10 000–12 000 (there are some 25 000 in the common English dictionaries).

It is the simplest words, those from within the 3000 word vocabulary, which most help a reader to receive a message. Writers who confuse communication by selection of complicated terminology and who attempt to display effulgent verbosity correctly acquire a corresponding reputation.

People don't read their rubbish.

USE OF VERBS
In science and engineering, there is a tradition of using verbs in the passive voice.

'Your drawings have been considered ... '

'A complaint was made by the clients.'

By contrast, the active voice conveys a higher sense of commitment and urgency.

'The client complained.'

Excessive use of the passive voice leads to a stodgy style.

CHOICE OF PRONOUN
Traditionally, engineers used to write in the third person.

'It is recommended that ... '

'In the opinion of the writer, it seems possible that ... '

The modern trend is to the more direct use of the first person pronoun:

'We recommend ... ' 'We suggest ... '

This choice between third person pronoun and first person is an important element of house-style, reflecting the character of an organisation. More conservative organisations will continue to use the third person. Those preferring to have an adventurous image adopt the first person.

There are then four key variables in writing style:

Length of statements: sentences, paragraphs, sections

Choice of words: keep it simple

Use of verbs: active voice generally preferred

Use of pronouns: first person or third

These are not the only variables, but if they are used sensitively, others tend to fall into place. For example, if sentences are deliberately kept short the writer will find that he is not using unnecessary adjectives; choosing simple words helps him to steer clear of abstract nouns.

In setting his style, the engineer has to steer a path between the prosaic and the effective. What should he take as standards for the way in which he uses the key variables?

Desirable style

The characteristic of writing for which the engineer should aim is that which will help the reader to follow him with interest, and indeed with enthusiasm.

That style depends on the reader, on his interests, knowledge, education, wants. Different readers will require different styles.

Different readers will require different styles at different times. The character of writing has changed dramatically during the past century, and the pattern of change has continued during the past decade.

The form of the change has been from extended, discursive writing style to shorter simpler construction. Some aspects of this change can be fairly precisely measured. These are shown, for example, by the reduction in the acceptable length of a sentence.

The writings of Dickens abound with sentences of up to 60 words: his average was about 32. Whereas a quarter of a century ago the average sentence length in *The Times* editorials was between 30 and 35 words, the corresponding length is now 26–28 words.

Nevil Shute, writing in the 1930s and 1940s, used an average of 28 words per sentence. Alistair MacLean, writing in the 1970s, used an average of 16.

After working with several hundred engineers concerned

71

with writing style, I suggest the following standards as being at present acceptable to professional readerships:

Sentence length. Average 26–28 words.

Paragraph length. Average three sentences per paragraph.

Choice of words: Use simple words. Include, for professionals, words within the normal professional language (but avoid one's own jargon when writing to professionals outside one's own specialism).

There is no simple measure which distinguishes simple words from more difficult words. However, longer words are generally in the latent vocabulary, rarely in the day-to-day. Most short words are in common use.

The engineer wanting to check his style can therefore check on the number of long words he uses. For this purpose 'short' means words with one or two syllables, 'long' means words of three or more syllables.

On this scale an acceptable level is that 12–14% of the words he uses can be 'long' ones. If he uses more, he is over-taxing both his vocabulary and that of his prospective reader.

Use of verbs: There is need for some balance between woolly use of the passive voice and the staccato use of the active voice. For a professional readership an acceptable style is found at a maximum of three verbs in every ten being used in the passive voice.

Tone and colour: There are three main aids to developing a desirable tone and colour in a style of writing. The first is to develop a style of writing which avoids the pitfalls suggested in the previous section. Competent writing style avoids causing difficulties for the reader. But of itself it does not make for an interesting style.

The second ingredient is the use of grammar. There are many well-recognised rules demanded in accurate use of grammar. For example:

Never start a sentence with a conjunction.

Never split infinitives.

There must be a verb in every sentence.

These are rules recommended for most situations.
But not all.

Vivid impact in writing can come from breaking such traditional rules of grammar. The statement 'But not all' is an unforgivable construction for a sentence, let alone for a paragraph. Yet it has a profound impact on the reader.

The engineer will not satisfy his readers if he misuses English grammar. However, it is possible deliberately to flout the rules of grammar to the reader's satisfaction and interest. If you are uncertain about the issue—consider your own reaction to the construction 'But not all' as a paragraph. Unpardonable? Or effective?

A third ingredient helping readers to be interested is the technique of 'orchestration'. The essence of the technique is, when a long passage is needed, follow it with a shorter one. It helps digestion. The break given by a short sentence both helps the reader's concentration and digestion and adds tone and colour to the writing. It is refreshing.

Tone and colour are thus based on writing style making helpful use of the key variables and by grammatical accuracy. They can be further helped by some manipulation of grammar and by the use of orchestration.

Use of layout

The writer has a battery of visual devices with which to tempt the reader. He must of course be discriminating in his choice within this battery. That which will be found attractive by one reader may be totally distasteful to the next. The battery of devices includes:

Changes of type-face. Even with a typed report, emphasis or variety can come from using different cases.

Underlining some words or passages.

Listing within paragraphs. Listing rather than going into prose.

Illustrating with photographs or documents.

Numbering sections, paragraphs, etcetera.

The way in which the writer uses this variety of aids should relate to the expected readership.

Different readerships

Different readers demand radically different treatment. They demand first of all that preparation should be reconsidered for each group.

A professional colleague—one who works at one's own level—is interested in what one does, why and how.

The boss is interested in what one does and why. If he becomes very interested in how one does it, then he delves into matters which are outside his personal scope: he becomes the sort of boss who does not delegate and who frustrates.

On the other hand, subordinates are interested in what one does and how one does it. For them, why one is doing it is a matter of remote and distant policy. It might be remotely interesting to know why one is so motivated, but it is not the stuff of their lives.

Such distinctions between readerships can be recognised for other categories. The one above relates to different authority levels, boss/peer/subordinate. Other distinctions include:

Professional and layman

One's own profession and other professions

The reader's level of experience

Native language English, second language English

Educational level

Technical level

Interest level

Just as different readerships demand different forms of preparation, so also they demand differences of writing style and layout.

For example, writing for colleagues or clients of a similar professional quality and experience, the engineer should use the recommended professional level—sentences averaging 27 words, etc.

But writing for a work-force on site, he has to get the interest of people who may not read any newspaper (or if any, a tabloid).

Here is a summary of the differences which professional writers use for the two categories. The first column of figures typifies the style in *The Times/The Guardian/The Daily Telegraph*; the second column, *The Sun/Daily Mirror/The Star*:

	Professional readership	Tabloid readership
Sentence length: average words per sentence	27	16
Paragraph length: sentences per paragraph	3.5	1.1
Use of verbs: percentage of verbs used in passive voice	30	1
Length of passage reader can be expected to digest (words per editorial)	750	225
Layout	Stark	Massive modulation
Type-face	—	Change of case. Italics, bold, stars, asterisks, underlining

In seeking to meet the different needs of his different readers, however, the engineer is circumscribed. He has to conform, within some limits, to that 'house-style' which is acceptable in the organisation which he represents; and he has to conform to the professional practices acceptable to his professional colleagues.

The outcome must therefore be a style blended both to house and to professional criteria. It must also be a reflection of his personal character. But always holding critical priority, he must remember the interests and need of his reader.

Economy of words

Readers do not like lengthy and verbose statements. Consider, for example, the following extract. It is taken from the first rough draft of a report which was to be sent from a satellite unit to superiors elsewhere:

'The Blank River Authority appointed Consultants because their own small staff resources were unable to undertake, in the time available, the extra work involved in schemes of this nature. If the Blank River Authority proposals are not accepted, alternative drainage schemes would be necessary to protect the motorway, involving the payment of additional fees to the Ministry's Consulting Engineers. In these circumstances, it is recommended that the Ministry contribute towards the Blank River Authority Consultant's fees; the cost being split on a pro rata basis to the work cost to each Authority. This would involve a Ministry contribution of £19 800 representing 54% of the total of £36 000. As will be seen from the correspondence and notes of meetings held, full discussions have taken place between representatives of the Ministry and the Blank River Authority, together with both Consultants, and agreement on apportionment is being sought as a result of these discussions.'

(151 words)

Compare that first draft with the second draft:

'In order to produce proposals promptly, the Blank River Authority has had to appoint consultants. If the Ministry were to proceed independently of the Authority, the Ministry would have to pay Consulting Engineers to produce alternative drainage schemes to protect the motorway. Accordingly, we recommend:

1. A contribution to BRA Consultants' fees for supervision of the project

2. Allocating these Consultancy fees pro rata to the Works costs

3. On those bases, a Ministry contribution of £19 800, being 54% of the total of £36 000.

These recommendations result from full discussion

between the Blank River Authority, the Ministry's representatives, and the Consultants.'

(98 words)

The reduction above, from 151 to 98 words, has been achieved simply by taking out unnecessary words. There is no significant reduction in the information being passed.

Such a reduction can often be achieved, even when the message has been clearly conceived.

There is always scope for further reductions. These depend on the amount of information which the particular reader may need. It might be, for example, that the 'superiors' to whom the particular report was addressed needed less information (or even more). This is a matter which depends on the writer's judgement of the reader's needs, and the scope is enormous.

Consider the possibilities:

'These proposals enable a sharing of Consultants' fees between the Ministry and the Blank River Authority. Such fees would otherwise have to be met in full by the Ministry. After thorough discussion, all parties have agreed to recommend sharing the Authority's Consultancy fees, pro rata to our respective Works costs. This requires a Ministry contribution of £19 800, being 54% of fees totalling £36 000.'

(63 words)

'Consultants' fees should be shared between the Ministry and Blank River Authority. By agreement with BRA, we recommend apportioning these fees pro rata to our respective Works costs. This requires a Ministry contribution of £19 800 (54% of the total of £36 000).'

(41 words)

'Consultants' fees:

1. These fees will be shared between Ministry and Authority.
2. Both parties agree on apportionment pro rata to Works costs.
3. The Ministry contribution will be £19 800 (54% of £36 000).

(30 words)

77

'Consultants' fees should be apportioned pro rata to Works costs. This will need a Ministry contribution of £19 800.

(18 words)

'Consultants' fees will need a Ministry contribution of £19 800.'

(9 words)

'Contribution to Consultancy fees: £19 800.'

(5 words)

The length of a statement can thus be reduced by 30% as in the opening example, by simple elimination of unnecessary words. Apart from such economy, length should be tailored to the amount of information needed by a particular reader.

Summary

The character of writing style may be judged by:

Length of statement (sentence, paragraph, section)

Choice of words: select the simple

Use of verbs: active voice encouraged

Use of pronouns

There are precise measures of these variables and standards which professional engineers find appropriate.

Tone and colour come from:

Choice of style

Grammatical accuracy/inaccuracy

Bridging

Orchestration

Layout can be used positively to command interest.

Unnecessary words may account for 30% of communication.

Different readerships need different treatments.

10

Letter-writing

Let us assume that the engineer has to write a letter. He has read the previous correspondence, checked and prepared calculations and information, considered the reader's needs, and has prepared the framework for the letter. What further steps should he take to send an interesting message?

Let us consider:

1. The reader's framework of interest

2. The personal element

The reader's framework of interest

The reader's interest will naturally follow that predictable path which we earlier discussed in Chapter 3. Initial interest, waning through the middle of the letter, reviving briefly at the end.

The structure of the letter therefore needs to take account of that pattern. It needs to capitalise on the opening phase of interest; to keep reviving interest through the middle; and to make use of the final burst of concentration.

For a start: alert the reader's mind to the subject area and give him suitable expectations. The title at the top of the letter should help him quickly to tune in to the subject matter. This is not achieved by a protracted title, such as:

'Drainage works at the eastern exit from the M6 motorway at the A34 (Stafford) intersection'

That title has two deficiencies. The impact is poor. It is a piece of continuous prose. There is much stronger impact if the width of title is restricted to one eye-span; that is, restricted to the width which a reader can see at one glance. That width, for the average professional, is three or four words.

The reader can better follow a sequence which starts with the general and then focuses increasingly on to specific detail.

An alternative way of treating the same subject matter would therefore be:

M6 motorway

A34 (Stafford) interchange

Eastern exit

Drainage

Short letters may require no more than a couple of pithy sentences after the title. Long letters, on the other hand, need suitable processes to help the reader's understanding and digestion.

Immediately after the title of a long letter the reader should be fed appropriate expectations of what is to come. He should be told the purpose of the letter and he should be helped to anticipate the way in which it will unfold. This will be in the form of headlines (signposts) showing the path to be taken. To take an example:

'This is to bring you recommendations about the drainage at this exit. The sequence of the letter is to consider:

1. Rainfall

2. Run-off

3. Catchment area

4. Outfall

5. Recommendations'

Through the middle of a letter the reader's attention should be regularly reinforced. Make generous use of sub-titling. It both refreshes by reminding the reader of words headlined in the opening paragraph and helps to add interest to the layout/appearance of the letter.

The practice of referring to a particular paragraph in previous correspondence is in general to be discouraged. Such a statement as:

> 'With reference to the comments in paragraph 6 of your letter of 14 June, the situation is ... '

This statement forces the reader to go back to that earlier letter, makes him use unnecessary time and energy. It is preferable to remind him of the text and then briefly feed him the precise reference so that he can check it if he feels the need.

> 'In response to your question about the capacity of the drains (your letter of 14 June, paragraph 6) the position is ... '

At the end of a long letter, capitalise on the final burst of interest. Summarise main points and particularly summarise any questions which have been asked or any actions which need to be taken.

After the letter has been typed, it is the writer's responsibility to check it for accuracy. He needs to check correct address, title spelling, wording. Great damage can be done by the accidental omission of the word 'not' in such a statement as: 'It is evident ... ' It is the writer's responsibility to check such detail. He may, of course, delegate that responsibility, particularly if he is fortunate enough to have a reliable secretary.

If delegating, he must of course ensure that they (writer and secretary) have discussed and agreed how they will share the responsibility for checking.

The personal element

The reader receives our letter in a pile with a dozen others. How can we influence the interest and attention which he gives to ours?

First impressions are crucial.

The very first impression is probably a subconscious one. He sees a letter-heading and recognises an organisation to whom he has some reaction, for better or for worse. That is inbuilt. The writer cannot influence that reaction.

He can, however, influence reaction in other ways. The reader reacts to what he sees in front of him. Here is a list of some ingredients which can have such a subconscious influence:

Spacing

Position on the sheet

Error-free typescript

Consistent margins

Does it look a mass? Or is it in digestible sections?

Do paragraphs look long and forbidding?

Does sub-titling break the mass?

Is there use of indentation to distinguish secondary from primary points?

Are sections and/or paragraphs numbered? (Some readers, in some circumstances, will be positive to this, others negative. It is a matter of taste and custom.)

These are all factors which will whet or spoil the reader's subconscious appetite.

His first conscious impression will be of the title and the opening words.

Hopefully the title will have the strong impact suggested in the previous section. The opening words, too, have a very strong influence. They need so to be chosen that they will have a positive influence for the particular reader.

I make much use of the words, 'This is to ... ' They are words which have a high impact; they are words which lead me naturally to specify the purpose for which I am writing the letter; and correspondingly they are words which lead quickly to the reader sharing my understanding of that purpose.

These same words may, however, have a negative impact. There are conservative readers who expect a more conventional opening. Nowadays the most conventional is to begin with the words, 'Thank you for your ... ' These are words with a lower impact. They do not capitalise so strongly on the opening potential to make a strong impression. Nevertheless, they are expected by some readers, and one then departs from the conventional at one's peril.

Through the middle of a letter the writer should be constantly awake to the niceties of the style he is using. He should be choosing a style suitable for a particular reader, as described in Chapter 9.

He should be refreshing the reader with his use of subtitling.

And he should be striving for 'reader orientation'. This is a subtle element of the character of writing. It is shown by some clear courtesies; for example, the practice of feeding a reader a brief summary to remind him of a point which had previously been made, rather than referring him back to the previous correspondence and expecting him to dig it out.

Another aspect of reader orientation is the way in which the writer uses pronouns. If he strongly uses the first person (I, me, my, we, us, our), he gives the appearance of being more concerned about himself than about his reader. If he makes excessive use of the third person ('It is recommended that ... '), his writing becomes woolly and stodgy. It is generally preferable to make more use of the second person (you, your, yours).

The positioning of pronouns is significant for their perceived strength. The first and the last are the most emphatic points of any statement. Thus the first word in a sentence is stronger than the same word used in the middle of a sentence.

Particularly strong is the first word in a paragraph, and outstandingly strong, the first word in the letter. Any letter in which the first word is 'I' ('We') carries with it an impression of an arrogant writer. In lesser degrees but nevertheless offensive is the practice of starting paragraphs with 'I' or 'We'.

There is only one circumstance in which a writer should make such strong use of the first person. That is the situation in which it is important for him to take on a mantle of authority in relation to the reader. The consultant writing to the contractor, for example.

The reader should be attracted to the letter by such sensitive attention to his interests and by careful use of writing style and layout possibilities through the middle of the letter. At the end, always leave with an element of goodwill. In a professional letter, express interest or readiness for the next phase. If the reader is well known personally, then end with a pleasant, possibly family, note.

Summary

The reader's frame of reference has in it the normal pattern of initial concentration, decreasing, rising again in the last paragraph.

Capitalise by starting with an over-view of the purpose and sequence of the message and by ending with a summary.

Ensure positive impact in the critical opening sentences.

Help concentration with breaks through the middle: sub-headings, possibly preceded by interim summaries and followed by section signposts.

Use layout as an ally.

Be sensitive, particularly about the priority between use of 'you' and 'I'.

End with goodwill.

I I

Report-writing

The writing of reports is, to many engineers, a difficult task. Not only is it difficult, but also it is very time-consuming.

This chapter is concerned with the writing of normal reports, possibly of 4–10 pages in length. Even more complicated problems arise in handling larger reports; these problems will be dealt with separately in the next chapter.

The framework for this chapter is:

1. Preparation

2. Report structure

3. Presentation of report.

Preparation

Most of us are familiar with the problems of writing reports. We start with subject matter which we know very well. We are, we believe, well able to write about it. It is clear in our thinking; we have done all the homework; there should be no problem.

Hours later, with several pages of scrap, we have ground to a halt. Despairingly we look back at what we have already done and decide that even that is unsatisfactory. So we start all over again.

Half the battle in avoiding this situation lies in our preparation. Unless we have systematically considered and written a plan for the report, the mind will be too full to segregate the sections. The brain will become overloaded, and we will inevitably despair.

The preparation method which I use follows that described in Chapter 2. First, the A4 process of clearing the mind of the clutter about the topic; then the systematic ordering (reader orientated) under a number of headings which the mind can readily contain.

It is worth taking plenty of time on this preparation, particularly at the stage of sorting one's thoughts systematically. Ten minutes' extra effort now can save a couple of hours' frustration in the later drafting stages.

There are some types of report which are produced repetitively. Such reports come to have structures with which both writer and reader are familiar. The sequence of thought and discussion is well known and therefore there is not any serious need to reduce to a very few headings. Some such familiar sequences are discussed later in this chapter and in appendices.

Despite the familiarity of such a sequence, the preparation stages are still needed to ensure that the contents within each section are coherently grouped; and within each section it remains desirable to sharpen thoughts, so that each section should be restricted to not more than four sub-sections.

In reports which do not conform to such standard sequences the preparation should ideally follow the hierarchy of four—a report in four parts, with four chapters of four sections . . .

The product of this preparation process is a skeleton of the whole report. When a report is going to be fairly long—more than say four pages—it will often be wise to do a further preparation stage before drafting each major section. Thus the process of producing the report might follow the sequence:

General preparation, 1 hour

Preparation of first section, $\frac{1}{2}$ hour

Drafting of first section, 1 hour

Preparation of second section, $\frac{1}{2}$ hour, etc.

In ordering his thinking, the writer should of course have the prospective reader clearly in mind. Not always easy, because reports are likely to go to a long distribution list, and also are often written with the thought that somebody may come along in future to criticise what has been done.

Nevertheless, it is important to have thought of 'the reader' during preparation. It is generally the case that a particular person (or type of person) is the real prospective reader. Others are on the distribution list either as a matter of courtesy or for information.

In general the reader should be regarded as the one who will have executive authority to implement recommendations in the report. The interests of his boss/bosses will be for a summary which should be given in an abstract. The executive himself will not wish to study all the technical detail but may want to have it checked by other people. The detail therefore should be given fully in appendices.

For effective preparation, therefore, consider the needs of the executive who will implement it.

Report structure

Many reports conform to established patterns. Specimen sequences can, for example, be given for:

1. Technical report
2. Structural investigations
3. Marketing report
4. Technical evaluation and recommendations

TECHNICAL REPORT STRUCTURE

When the engineer has carried out a technical investigation he needs to present his findings.

Conceptually, he should have in mind four stages for a report. They answer the questions:

1. What is the problem?
2. What have we done about it?
3. What does this show?
4. So what?

These questions come into a standard sequence described below.

That standard sequence, the main body of the report, should be topped by an abstract and by a list of contents, and should be followed by appendices. Before dealing with these surrounding sections, let us first consider the main body of the report.

Introduction
This should cover:

> Purpose of report
> Scope
> Shape

Statement of problem
An accurate definition of the problem is often the most important step in the conduct of an investigation, and it should be stated clearly at the outset. Even when the problem was clearly defined and agreed from the outset with the client, the statement should be included in the introduction.

There is also merit in including at this early stage of a report a facsimile of a client's letter of instruction, commissioning the project and report.

Acknowledgements
There is good psychology in acknowledgements to other parties being given at an early stage of the report. There is of course particular advantage when acknowledgements are being made to client's staff.

References
This should be a straightforward list of references (to books, articles, etc.) referred to in the report. The case for the positioning of references here in the text is only marginally stronger than the case for relegating them to an appendix.

Literature survey
If there is a wealth of literature there is a temptation to quote a great deal in a report. The temptation should be resisted, bearing in mind the real interests of the reader. Give him the essence of the literature, summarised. Extended statements of the literature, if needed, should be included in appendices and not in the main body of the report.

Methods used

State standard methods used. Outline any special methods developed for the special case. Again, keep it brief, using appendices if long written statements are needed.

Problems encountered

The reader should be made aware of any special difficulties or hazards which were met. It is best to do so at this stage. It is wrong to leave them to a later stage, when such problems outside the writer's control might be seen to be confused with his comments and conclusions.

Summary of results

A summary only. The essence of what the executive reader needs to know. The detail should be included in appendices so that his staff have a chance to check.

Comments

Up to this point the report has been entirely objective. The engineer has presented his understanding of the problem, what he has done and the clear results of what he has done. Now, for the first time, comes the point at which he applies his subjective judgement. He states how he values the results.

Conclusions

These should follow clearly from the logic of the work done. Ideally, the reader is recognising the implications even before he reads them.

Recommendations

Finally, the writer commits himself to the action which, he believes, should follow.

Covering the main body of the report should be an abstract and a list of contents. It is not the primary purpose of the abstract to inform 'the responsible executive', at whom the main body of the report is aimed. The abstract is for that executive's boss, for his boss's boss, and for others who need to know what is going on without having time to absorb the full report.

The abstract, which may well be circulated separately, should cover material in three sections of the report:

Introduction (purpose and scope of report)

Conclusion

Recommendations

Subject only to some paraphrasing (and to omission of 'shape of report' from the introduction), it may be possible simply to reproduce those sections as the abstract.

Appendices will be of four types.

1. *Statements* too long to be included in the main body of the report.

2. *Tables and figures.* The advice for writers producing such tables is necessarily banal. Ensure all tables are numbered and titled. Ensure that all columns are clearly headed and that units of measurement are defined. Ensure that all the figures quoted are well checked.

   The advice is banal but in practice these are details which we all tend to overlook, especially in the heat of the last rush of getting out a report. It is prudent always to have a third party who will check that we have got these details accurate.

3. *Diagrams.* Similar caveats apply. Ensure that ordinates and axes are clearly defined. Do not cover diagrams with too much information. The reader is more helped by several simple diagrams than by cluttering the same data on to one complex diagram.

4. *Photographs and maps.* Where the reader could be helped by frequent reference to a photograph or map, it should be presented on an extended sheet so that he can see it open alongside the text while he is reading.

What about authorship? Should the author's name appear? While this is not common in engineering practice, it is common in the world of science.

And what about distribution lists? Should they be printed? Again it is not common practice in engineering, but it is a practice which can be helpful. The client may be helped if he knows who else in his organisation has received a copy. And the engineer's organisation should have a separate list showing distribution of internal copies.

In addition to such standard structures for a technical report, there are repeatable structures for other forms of report. Three of them are presented in appendices:

Appendix 1 Report format for structural investigations

Appendix 2 Marketing report

Appendix 3 Standard structure—technical report

In the standard structure suggested the sequence becomes a familiar one. It is not always necessary, of course, to use every heading. In a technical report the problem may be adequately stated within the introduction; there may have been no literature survey; there may be no need for some other sections. But having a standard sequence from which to select is helpful to both writer and reader.

Presentation of report

The writer wanting his report to be read with interest must not only prepare a structure which the reader will find logical and easy to follow.

He must also develop the manner in which he presents information. The reader is human and will form a liking or a dislike for any report. That liking will of course be influenced by the content—by what is said. It will be influenced too by how it is said. It is partly an emotional matter, not entirely logical.

This human element will be considered below in four main sections:

1. Introduction
2. Bridging
3. Style
4. Indexing

INTRODUCTION
As ever, first impressions are crucial. The report reader forms

these opening impressions on what he quickly sees of the outside and the opening pages of the report.

First there is the external appearance. He should see something which is attractive. Attractively bound, in an attractive colour, in impressive material, with wording restricted to essentials: a succinct title in a substantial type-face; the client's name at the head of the sheet, slightly smaller, and the writer's organisation at the foot, smallest of the three.

Then there is the internal appearance, which will influence the reader in five ways:

1. Sheer bulk. A great big volume is a disincentive to any reader. Help him to see how little it is essential for him to read. Distinguish that fraction which is the main body, from the mass of appendices. Distinguish either with different colours of paper or by use of distinctive and obvious dividers.

2. Layout can make or mar a first impression. A stodgy mass on a page is repulsive. A neat, well-broken, well-spaced appearance, with appropriate use of indentations and sub-titling, can make it attractive.

3. The opening sequence is important. The reader opens the report without knowing what is to come. Sometimes he is catapulted into the text without any idea of how it is to develop. It is far better if the writer ensures that the reader quickly finds his way around the report. The recommended sequence is abstract, followed by contents and then the introduction to the main body.

4. The choice of opening words and their composition should have strong impact. It is worth spending double time on the opening sentences. Ensure that they have high impact; then transfer that impact to the abstract.

5. The content of the introduction is important. Crisply and clearly feed the reader the purpose, scope and shape of the report.

BRIDGING

In the main body of the report the reader is required to digest a lot of information. His digestion is helped by feeding him appetising dressings from time to time. In particular it is helped by the technique of bridging.

This is a technique of giving him, at the beginning of each section, a taste of what is to come. Then at the end helping his digestion with a reminder of the essentials.

This is distinct from so many engineering reports in which two or three pages of facts—for example about hydraulics—are presented in a solid stodgy sequence, ending with number 3.14; turning over the page and seeing '4. Mechanics' and a further stodge numbered '4.1' and running on over two or three pages more to '4.17'.

Use the bridging technique to help the reader's movement from one section of territory to another. Open a section with words preparing his mind for what is to follow. For example:

4. Investigation phase
The key parameters studied in the investigation stage were:

1. Water

2. Tides and surges

3. Temperature

4. Identification of other hazards

Develop the section, presenting the facts which emerged; then move on to the bridging from this section to the next. Summarise key points from the former; then follow the next sub-title with an introductory paragraph for the section. For example:

The main facts shown by the investigation are therefore:

1. The water is highly saline.

2. There is little variation in tides, but there have been surges of up to 6.3 metres.

3. There are diurnal temperature variations of 25°C and seasonal variations of a further 20°C.

4. No other hazards have been identified.

Those are the facts.

Discussion of results

Key issues in interpreting the facts are:

(a) ...

(b) ...

(c) ...

(d) ...

Such a bridging technique is helpful to the reader, both from section to section and from chapter to chapter.

STYLE

The main elements of writing style have already been discussed in Chapter 9.

Within a report, the style should of course be appropriate for the reader. If writing for a professional audience, averaging sentences of 28–30 words, three or four sentences per paragraph, a vocabulary of up to 6000 words. Unless the reader is from the same discipline as the writer, careful avoidance of jargon.

If writing for overseas clients who do not have English as a native language, a simpler style and use of a smaller vocabulary.

If writing for a less educated readership, then again a simpler style: possibly 20 word sentences, single sentence paragraphs, 3000 word vocabulary, continuously active English.

INDEXING

The use of indexing—attaching numbers to chapters, sections, even paragraphs—has two benefits.

First, it enables the reader and the writer more easily to communicate with one another when they later want to refer to a particular part of the report.

Second, it helps the reader to recognise the significance of each section or paragraph.

This is particularly helpful when the writer uses a multi-numeral system. For example, the first number referring to chapters, the second to sections within a chapter and the third to paragraphs within each section.

Thus the figure 3.2.4 is paragraph 4, in section 2 of chapter 3.

There are thus strong benefits to be had from using an indexing system. There are also two disadvantages.

First, some readers are irritated. They feel that an excess of numbering is patronising.

Second, the writer may be tempted to use his numbering system in early drafts. He then tends to become a captive of his own system. His numbers restrict his thinking and his flexibility in re-drafting.

My choice for reports, balancing the advantages and disadvantages, is to number chapters and sections, but not to go any deeper.

Summary

Key issues in report writing are thus:

1. Preparation:

> Orientated to the executive who will have responsibility for the implementation

> Thought processes organised systematically to provide the skeleton of the report

> More detailed skeletons for long sections

> Structures where possible conforming to recognised sequences

2. Structure:

> Standard structures help reader and writer

> Novel structures geared to the 'four main points' concept

> Abstract aimed at 'responsible executive's superiors'; a paraphrase of introduction, conclusion and recommendations.

3. Presentation of report:

> First impressions crucial: appearance both external and internal; opening content and style for critical impact.

> Bridging as an aid to the reader's understanding

The style chosen to suit the executive responsible for implementation

Limited use of numeric references

These then are important issues in the writing of most reports.

12

Major reports

There are further problems in handling major reports.

We shall in this chapter be concerned with the sort of report which is an inch thick or more. It has taken weeks or even months to prepare. Several departments have probably been involved. It may be the only visible evidence to the client of his expenditure and of the engineer's professional competence and effort. It is the sort of report that regularly gives writers ulcers.

The sections in this chapter deal with:

1. Organisation
2. Preparation
3. Drafting
4. Scheduling

Organisation

The production of such a major report, involving team effort, needs a special project organisation.

Important aspects of this organisation are:

Leadership. Identification of the individual who will have responsibility and authority for the preparation of the report.

Issuing authority. If different, the individual who will take responsibility for the issue of the report.

Report control group. Key people who will control contributions to the report. Ideally, such a report control group should consist of the group leader, plus representatives (preferably heads) of all departments who will make a contribution.

Planning and progress control arrangements. Meetings of the group at sufficiently frequent intervals to monitor progress and sustain impetus.

Schedule. A statement of what needs to be done by when. Suggestions for preparing this schedule will be found in Section 4 of this chapter.

Preparation

In principle, the technique for preparing a major report follows the A4, A5 approach discussed in Chapter 8.

In practice, for a major report, the technique needs to be modified.

The technique should now be used in at least two phases. *The first phase* is aimed at producing a structure for the report as a whole: the parts and the chapters in which it will be presented, possibly going as far as sections. *Second phase preparation* is at the more detailed level of developing the structure for each chapter, with its sections and with the main material to be included in each section.

Preparation for the report as a whole should be done by the report control group. They should have a preparation meeting or meetings at which they should develop the skeleton in three stages.

Stage 1 is the brainstorming stage, the A4 stage. The group—ideally working with a flip-over chart—brainstorms the full range of ideas in their minds as to what might be included in their report. Given a lively professional group, the thoughts of each member are stimulating to the others and a big flow of ideas can be expected. Certainly more than in the two-minute framework which is possible for the individual working on his own.

The brainstorming needs to be free flowing: there must be no criticism of ideas, no demands for justification, no deep diving as one topic after another is articulated on to the flip-chart.

Stage 2 is the thesis sentence. This is a definition of the writer's real purpose in writing the report. What is it for? What is it *really* for?

This purpose may be different from that purpose which will be declared in the report itself.

The thesis sentence must be unanimously agreed by the control group. This is not easy. When eminent professionals sit together at this stage there may be hidden differences of view about the real purpose of the report. In the urge to get on with today's business, those niggling differences can all too easily be swept under the carpet. Unfortunately, they have a habit of emerging again two months later after contributing to the ulcers that the writers have been developing.

It is important to exorcise the differences of view at the start even if it takes an hour or so. Be glad if it does. It means that the group has already recognised problems which would otherwise have been very costly in both time and temper at a later stage.

Having at last produced a thesis sentence, subject it to one test. Is it longer than 20 words? If it is longer, then the group must continue discussion until they have a sentence of the simplicity of 20 words.

The key result of this process is not so much the 20 word sentence which is produced. It is rather in the extent to which the key people have been through the process of sorting out their ideas. It might well be that the same sentence might be interpreted differently by other people; but as long as the principals have been through this procedure, then they can control and co-ordinate the report and work together on it.

Stage 3 is the A5 stage of structuring the report.

Normal guidelines apply. Do not try to put in everything. Think of the reader, his interests, the headings which he will find helpful. Unless the framework is a familiar one, regularly repeated in many reports, try to build up around the 'hierarchy of four'—four parts each with four chapters, each with four sections.

This A5 stage of structuring may very well be time consuming. It deserves to be. One hour of additional effort put into getting the structure right at this stage can save a week of re-drafting at a later stage.

Beyond this three-phase preparation—A4—thesis sentence

—A5—the control group may be able to go a stage further. They may for instance be able to suggest a consistent theme for the internal sequence within successive chapters or sections. Here are some suggestions for internal sequences, using different themes:

Problem-solving sequence

>   What is the problem?
>   What are the causes?
>   How important is each?
>   What are the options?
>   What is the best?

Instructional sequence

>   What to do
>   How to do it
>   Why

Professional discipline

>   Chemistry
>   Physics
>   Geology
>   Geography
>   Economics

Chronological

>   What happened
>   Present position
>   What next

In the first phase of preparation the control group will have established a structure for the report as a whole.

Each member now goes back to his own department. Within the department, there is now the need to plan the department's own contribution to the report. This should mirror the three-stage approach advocated for the control group, the department's contribution being established through its group-work of A4—thesis sentence—A5.

A great deal of work will by now have gone into the conception of the report. But this effort pales into insignificance when it comes to its birth and development.

Drafting

The birth process is laborious. Typically it causes great agony to the writer. He sits down, starts writing, gets a few pages done, gets called away to something else. He goes back, picks up the pen, re-reads the first few pages, finds they are dreadful and re-writes them.

Then he finds it difficult to progress through to the next sections. He burns a lot of midnight oil until finally he produces something which he takes to the issuing authority. He is ticked off for splitting his infinitives and for not getting sufficient orderliness into the presentation. He goes away sad and confused, possibly rebellious.

He and the other members of his work-group—including his boss—need a discipline which will enable them to overcome some of the difficulties that he is meeting. This can be found by having a phased approach to the drafting of a report.

I use a four-stage drafting process. There are four drafts, each serving a different purpose, each on a particular colour of paper; pink, blue, yellow and white.

The pink draft is the first. The purpose is to get something on to paper. Anything which will act as a basis for review, criticism, development.

It is always grossly imperfect.

Its purpose is to enable me to check whether my structure is sound, and to begin reviewing the content.

Different people have different ways of tackling this drafting stage. Most people apparently like to write with pen on paper. Personally I find this not only laborious but also positively difficult. My mind gets captured by the details of the line I am writing; and physically the process is such that I regularly go back to the beginning and re-start. This is a real waste of time at this stage when one should be seeking for the roughest of products. My method is to dictate, starting on page 1 and going straight on; and I have a contract with my secretary that I will not be shown page 1 before I have dictated the last page.

There is some limit to one's capacity to concentrate and be efficient in this writing process. My experience is that during one day, my competence is to achieve second-phase preparation for 4000 words and then to dictate that amount. This takes me about $1\frac{1}{2}$ hours for preparation and $2\frac{1}{2}$ hours for dictation.

The outcome of this process is a pink draft. This should show that the gross structure (of the main parts of the report) is satisfactory, and that the scope of most of the chapters is also satisfactory.

The draft will generally show, however, some need to change the sequence of chapters, to prune one or two, to divide one or two, to enlarge a couple. One or two will be so abominably composed that a second pink draft of them is needed. Others will read more happily, and may even be found satisfactory on a section-by-section basis.

One outcome of using this disciplined approach to preparation is that the writer at this stage has produced a deliberately rough draft. It is expected to be full of errors and *non sequiturs*. That is the natural order of things and he has little need to feel defensive. He is able objectively to join in discussion and criticism of his draft.

In reviewing, he and his colleagues might take the following agenda:

1. Recapitulation:

   Prospective reader

   Thesis sentence

2. Consider:

   Scope of report

   Size

   Gross structure/shape

   Sequence of chapters

   Scope of chapters

   Consistency of internal sequence

The draft will need severe re-writing. It is therefore a waste of time at this stage to consider:

Paragraph by paragraph

Writing style

Grammatical accuracy

Choice of words

The pink draft is thus a first shot, very rough.

The blue draft aims to marshal the content, to get it developed and balanced.

Much of the first draft will need a complete re-write. My experience is that if I try to dictate afresh with the original draft in front of me, I become constricted within the thought-pattern of that earlier inadequate draft. To free my mind to think creatively, I need to go back to the stage of recreating a phase 2 A5: that is, I need to recreate the skeleton on a chapter-by-chapter basis.

The outcome will be a blue draft which, it is to be hoped, is satisfactory not only on a chapter-by-chapter basis but also on a section-by-section, or page-by-page basis. There is now much which will be retained in later drafts, and organisations with word processors may find it worthwhile using them for the blue draft.

Discussion of the blue draft should focus on content, and *not* on style. It is a waste of time to criticise style before the content is adequate.

The yellow draft is the stage to be concerned about style. Check for grammar, check for orchestration, check for bridging, check for jargon. Check and re-check the impact of the introduction.

The final phase is the white draft, the completed final draft, visually appealing, accurate. Discuss beforehand with the production experts, the secretary, the printer, the binder. Make imaginative use of layout and type-face possibilities.

Have a critical colleague check through the appendices, looking critically at the detail. Consider presentation. The external appearance, the internal distinction between main body and appendices, and especially the critical first impressions.

To summarise: this section has suggested a disciplined

approach to the drafting of reports. It has suggested a four-stage approach:

The pink draft to get the skeleton right

The blue draft to get the content right

The yellow draft to get the style right

The white draft to get the presentation right

The discipline is a tough one to follow. Writers naturally worry when they see for themselves that an early draft is illogical. Both they and their critics can all too easily be magnetised into criticising and amending paragraphs and grammar, and even words, at all too early a stage. It is wasteful of time and energy to do so. The disciplined approach, both for writer and for his superiors and colleagues, pays dividends.

Scheduling

It is always difficult to schedule accurately the preparation of a major report. Invariably, the contributors begin with the belief that they know all the answers and make assumptions which later prove to have been grossly optimistic. The report preparation runs into all sorts of difficulties, both the difficulties of creative writing and the difficulties of liaising between departments.

Much midnight oil is then burnt and much worry created in trying to achieve schedules which were never really possible. Failure means that the client gets the report late, and even then that an imperfect draft is sent to avoid even further delay.

There are two ways of averting this trauma. One is the traditional:

'Think it out very carefully, making generous allowances. Then whatever the answer is, multiply it by four.'

The alternative is to seek a more logical approach.

The time which will be needed for drafting will vary from organisation to organisation and will depend on the extent to which drawings have previously been prepared, the need for additional materials, the extent to which there has to be dis-

cussion and revision between different authors and departments.

As a very rough yardstick, it is my experience that at each
drafting stage I need one hour per 1000 words to prepare and
construct or reconstruct.

I also find that concentration sags if one concentrates for a
whole day on writing—and anyway it is impracticable since
we all have other demands on our energy and cannot commit
ourselves utterly to writing.

My unit of measurement for scheduling is therefore 'half
days'.

On that basis, a schedule for a report of say 10 000 words
would be:

	Half days
Preliminary discussion and preparation	2
Secondary preparation and dictation of pink draft	3
Colleagues' reading time and discussion	2
Blue draft	3
Discussions	2
Yellow draft	3
Discussions	2
White draft	3
Total	20

On the basis that the writer would commit himself for half
his time to production of the report, the total production time
would be 10 days (one day per 1000 words), spread through
four working weeks.

This is, however, a very rough yardstick. Each organisation
will have different experience and should from its records be
able to analyse its own scheduling needs.

Summary

The production of a major report is the stuff of which ulcers
are made.

Writing this major report is made, not easy, but less ulcer-
ous if the writer has a systematic approach.

The recommended approach is:

Organisation

    Clarify leadership
    Clarify responsibilities
    Form steering group
    Define planning and progress arrangements

Preparation

    Brainstorming
    Thesis sentence
    Skeleton

Drafting

    Pink for structure
    Blue for content
    Yellow for style
    White for final

Scheduling

As a very rough yardstick, the full drafting process might be scheduled at one full working day per 1000 words.

Allowing for a writer committing half his time to other duties, anticipate completion in two days per 1000 words.

13

Reading skills

Engineers need to read a great deal, and their effectiveness depends on their ability to digest this material adequately and time-effectively. That is what this chapter is about.

There are three distinguishable types of reading.

First, there is precise reading. The engineer needs to study the precise figures, to meditate on them, to criticise them, to see their implications. In this form of reading, the actual reading time is subordinate to the time during which he is cerebrating—and that is outside the scope of this book.

A different form of reading is that which he does for relaxation. The novels he reads. The 'hobby' magazines. His objective now is relaxation, winding down, getting away from it. The style of that reading will have an influence on his own writing style, and to that extent it is preferable for him to read good quality books, journals and newspapers. But the way he reads them is not important. What matters is that he should relax while he is reading them.

Between these extremes of precise reading and relaxation reading, there is a middle ground of picking up information which he needs for his business. The potential volume of this information is infinite. What now matters is the efficiency with which he uses the time he can reasonably devote to reading.

That is what this chapter is about. It suggests two techniques which enable the reader quickly to get the essence of information of professional quality.

One such technique is concerned with getting the essence of a major general work such as a book. The other is a technique for rapid reading of any passage, maybe an article, for example.

Key learning

Suppose the engineer needs to read a 'general information'-type book. His normal reading time, from starting page 1 to ending at page 340, might be 10 hours.

But there are only a few aspects of the full book which will usually be of special interest to him. There are some things which matter a lot; there are other things which do not.

The recommended approach for this situation is a three-phase approach:

1. Quickly establish the scope and shape of the book.
2. Formulate a few questions which the book should be able to answer.
3. Search for the answers to those questions and only those questions.

In more detail: in phase one, sit down with a clean sheet of paper. Open the book in the middle. See what the left-hand page is about. Jot down a note of anything stimulating. Move on another 40 pages. Repeat.

Go back to the contents page. Where are the bulk of the pages taken up? What looks to be of most interest? Open the book again a quarter of the way through, and at random at other pages. Is there an index? What are the items with a lot of entries? What is interesting?

What is promised by the blurb on the cover? The author's own foreword? What of that seems important? Constantly jot reminders, seek only to familiarise with the theme and form of the book—not yet to understand what is written.

Allow 12 minutes for this opening phase. The product will be, in one sense, a sheet covered with jottings. In another sense the product will be that the reader has a general idea of the scope and the shape of the book.

In the second phase sit back and consider: what is this book about? What are the interesting things in it for me? What are the four most important questions I could get answered?

Next, on another sheet of paper, write down the title of the book and four side-headings, one for each of the four questions which one would most like to get answered by the book.

This second 'thinking' phase, of establishing the main questions, should take some five minutes. The product is a skeleton comparable to the beginnings of an 'A5' stage of preparation.

Now start again the process of random dipping into the book, looking out for material relevant to the key questions. Ignore everything else. If within half a minute of opening one page you have found nothing relevant, switch quickly. Jot memory ticklers of relevant material on a fresh sheet of paper for each important question.

After about 10 minutes of this further random diving, backed by the experience from the first phase of the exercise, the reader will find that he knows his way about the book reasonably well. He can now become selective, using his knowledge of the shape of the book in combination with the evidence in the contents list and the index.

His objective now should be to exhaust the answers which the book holds for the questions which are of the most importance to him.

Continue this stage for about half-an-hour.

Finally, revert to the A5 and write under each question a synopsis of the answers in so far as the book provides them.

This whole process has taken about one hour. The engineer who has followed the process will find that he is confident that he knows the sort of answers which the book can provide for him. Not only the answers to the questions he posed, but also the author's pattern of thinking on other issues.

He will in fact have imbided 90% of the information which he could get from this ten-hour book, in a one-hour period.

The process, to repeat, has the stages:

1. Random jottings

2. Frame key questions

3. Dip into book again, at first randomly but becoming increasingly deliberate, to seek the book's answers to the questions.

Rapid reading technique

The typical engineer reads at a speed of some 200–250 words per minute. However, some people have an ability to read much faster. They claim to be able to read a paper-back between Liverpool Street and Bishops Stortford—and some of them are indeed people who 'read' a book at this speed. They are exceptionally gifted.

For the rest of us, such a remarkable speed is outside our potential, but we do have the possibility of increasing our reading speed three-fold or four-fold.

There is evidence that when we do improve our reading speeds in this way, our comprehension does not suffer. The tests show in fact that we will retain about 70% of the material which we read at (say) 200 words a minute but nearer 80% when we increase our reading speeds to about 800 words a minute.

What is the technique? There are four key steps:

1. Concentrate. Get the full energy on to the material being read and sustain that level of concentration.

2. Avoid regression. Never never go back to study a passage only half digested. This discipline has a powerful influence in helping concentration during first-time reading.

   To implement take a pen or ruler, lay it along the line above that which you are reading and keep the ruler moving consistently down the page. Never let the ruler go back up again.

3. Avoid sub-vocalising. The slow reader tends to enunciate to himself (sub-vocalise) each word that he reads. His speaking speed is probably under 150 words a minute, but his potential reading speed is probably 800. Sub-vocalising destroys the possibility.

4. Maximise eye-span. Each of us has an ability to take in a certain breadth of text at one fix of the eye. A bad reader will fix on each of eight or ten words in a line of text; a slightly better reader will capture two words at each span; a competent engineer has the potential to capture three or four words at each eye fix.

To develop this expertise, he should make a habit of maximising eye-span. When reading newspapers, which typically have five or six words per line, the reader should aim to capture each line with one eye fix. This means that his ruler will go rapidly down the column. His eyes will capture the central three or four words in each line. The edges, possibly one word at the left-hand end and one at the right hand, will be blurred but this will not prevent him getting the sense of the message.

With concentration, avoidance of regression, avoidance of sub-vocalisation, and this pattern of maximising eye-span he will find that his understanding of a newspaper article is high, possibly even enhanced while he doubles or even quadruples his reading speed.

The corresponding technique for reading books, with their longer lines, is to restrict eye fixes to two or perhaps three per line.

Summary

This chapter has suggested two techniques to help reading efficiency.

Key learning technique:

> Random jottings
>
> Frame key questions
>
> Dip into book again at first randomly but with increasing deliberation
>
> Jot reminders of all answers to key questions

Rapid reading technique:

> Concentrate
>
> Avoid regression
>
> Avoid sub-vocalisation
>
> Maximise eye-span

Part 3

Effective meetings

Introduction to Part 3

This part of the book is intended to improve effectiveness of meetings.

The main focus is on business meetings where the number of members is likely to be between 4 and 12 and where the normal purpose is decision-making.

Much of the effectiveness of business meetings depends on the effectiveness of the chairman.

His task is covered in three chapters:

> The first deals with his conduct outside the meeting. The second with his control of meeting procedures. The third with his control of the members.

Those members should be effective in their contributions and we devote two chapters to those contributions:

> The first to the presentation of cases. The second to members' behaviour at other times during the meeting.

Three further chapters complete this section on meetings. They are concerned respectively with:

1. The role of the secretary
2. The life-cycle of committee meetings
3. The role of the chairman and of members at larger gatherings and conferences

14

Chairmanship
(1) Outside the meeting

The chairman's effectiveness depends on his conduct of a meeting and that conduct can be made or marred by the work he has done outside the meeting. It is discussed in sections dealing with:

1. Administrative routines in which the chairman and the secretary have roles to play
2. The chairman's discipline for planning the discussion on each item of the agenda
3. Physical arrangement of the meeting room
4. Action after the meeting

Administrative routines

It is the chairman's job—possibly in consultation with, or even delegated to the secretary—to ensure that dates for meetings are set, venues reserved, and members notified. Not least of his responsibilities is to check on the purpose that would be served by holding a meeting. All too often, a committee gets into the habit of meeting at regular intervals and goes on meeting at those intervals long after its original purpose has been served. More on this in a later chapter on the life of committees.

The chairman should ensure that members are given the chance to suggest agenda items, and it is his responsibility to approve the agenda. He should ensure that papers for a meeting are sent to members in good time. There is no hope of running an effective meeting if members haven't had the chance to read and digest the documentation.

It is possible for a chairman to manipulate the agenda. For example, it is easy for him at an early stage of a meeting to take a protracted argument on a trivial item and then at some later stage to slip an important point through at the end of 'other business', while members are getting their papers together ready for departure. In my experience, this is an important part of the armoury of power in universities and in bureaucratic organisations. In business organisations, it is suspect and likely to rebound to the discredit of the manipulator.

There is then a range of mechanical routine responsibilities which the chairman must either discharge himself or make sure are discharged by the secretary.

Planning the discussion

While the chairman may delegate the administration, it is imperative that he himself should plan the discussion.

The purpose of his planning should be to create a path which enables the members to contribute positively, helps them to progress rapidly and smoothly and stops them from chasing red herrings.

The method here to be recommended for planning the discussion follows, in principle, the same three stages as the approach in Chapter 2: an approach with first of all a random 'ideas' phase, followed by an analytical phase, and ending with some simple aid for use in the meeting. This three-phase process should be done for each item on the agenda.

The first phase is the normal A4 phase. A short period of clearing the mind, putting on to paper in random order the jumble of thoughts about the topic. An example appears in Figure 14.1.

The second phase is the analytical, and the chairman should direct his thinking to three distinct topics in this phase.

First, the purpose of the discussion. Is it, for example, to decide a new policy? Or is it to decide how to implement a policy previously agreed?

All too often, discussion on a controversial subject is

Purchase of steel

Progress review

Quality requirements	Short-list
Tenders received	Length ?
	Who ?
Progress on elimination	Priorities for negotiation
Germans' reservations	Problems
Japanese shipment	Risks — especially strikes
Price differences	Co-ordination of suppliers
Quality assurance	Co-operation with contractors
Get Jack's comments	Responsibilities for testing
Excess guarantees	Ensuring full government support
Banker's agreements	Money
	Technical
Financing	Safety
Export credits	Revised technical specification
Negotiating team	Commercial evaluation
Technical members ?	Price
	Discount
Team training	Terms
Negotiating dates —	
Start ?	
Finish ?	
Turnkey	
Delivery promises	

Figure 14.1 The first stage of chairman's preparation—the A4

unnecessarily protracted. The decision needed may be on whether to take some course of action—for example, 'whether to tender for Southport Sewage Works'—yet the discussion rambles deeply into how such a project would be implemented.

119

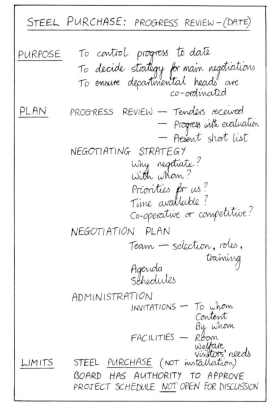

Figure 14.2 The second stage of chairman's preparation—the A5

Eventually a decision is taken. Then at their next meeting a month later, when they should be discussing progress with the tender, the members find they are going over all the same arguments as before, on whether they should be tendering at all.

This is a pure waste of time. Everybody knows it. It happens time and again because the chairman has not thought out the purpose for which this item is on the agenda.

The competent chairman distinguishes between:

> *Policy* decisions, and decisions on *implementation* of policy previously agreed
>
> *Strategies* and *tactics*
>
> *Planning* and *control*

STEEL   PURCHASE
PROGRESS   REVIEW

PURPOSE   CONTROL PROGRESS
                 DECIDE STRATEGY
                 CO-ORDINATE

PLAN   PROGRESS REVIEW
                 NEGOTIATING STRATEGY
                 NEGOTIATION PLAN
                 ADMINISTRATION

LIMITS   NOT INSTALLATION
                 PROJECT SCHEDULE
                 BOARD'S AUTHORITY

Figure 14.3 The third stage of preparation—postcard simplicity

Second in the analytical part of his preparation, the chairman should prepare a plan for the discussion.

Rather than allow discussion to wander all over the place on each item of the agenda, the chairman should have a clear plan, a sequence of landmarks which will enable him to keep the discussion moving forward within reasonable bounds.

The ideal number of such landmarks for a discussion is four—the number which both chairman and members can keep sharply in mind.

For example, having agreed at a previous meeting that we should tender for Southport Sewage Works, at our next meeting it could be that the purpose is to agree how the tender will be made. It might be, for example, that the headings would then be:

(a) How keenly to tender

(b) Possibilities of extra work

(c) Staffing

(d) Schedule for preparation of tender

(The word 'schedule' is here used in the sense of what needs to be done, by whom and when.)

The chairman also does well at this analytical stage to consider limits for the discussion. For any item there are boundary issues, grey areas which are of marginal relevance to the topic. Meetings all too often can get bogged down in and beyond such marginal issues—occasionally because of deliberate manipulation by members anxious to voice their self interests, more often unwittingly, as the momentum of a meeting takes it outside reasonable limits.

Such time-wasting can be forestalled if the chairman has done his homework properly and has anticipated likely red herrings. For example, in the discussion of how to tender for Southport Sewage Works some of the red herrings might be:

Progress with Northend Sewage Works

Recent troubles with nominated sub-contractors

The need for more staff in the Estimating Department

The product of the analytical phase of the chairman's preparation for each agenda item should thus be the A5 with its definition of purpose, plan and limits. See the example, Figure 14.2.

The final phase of this preparation is to produce the chairman's aide-mémoire for use in the meeting. The A5 contains too much information to be sharply visible. He needs something much sharper to help him keep control in the hurly-burly of a meeting. That sharp something is ideally a postcard, with one word printed large for each aspect of his plan:

One word for agenda item

One word for purpose

One word for each of his main headings

One word for each of his limits

An example of this A6 phase is Figure 14.3.

He should also be considering how the discussion is to be stimulated. In general, it is helpful to identify an individual who is prepared to open the discussion on each item. The chairman might take the initiative by contacting a colleague and asking him to lead on a point. Alternatively, a member can be nominated on the agenda papers. If he does so decide, he should put the member's initials on his postcard.

The product from such chairman's preparation is two-fold.

One product is his aide-mémoire, a series of postcards, one covering each item on the agenda. The other product is that his mind is alerted and sharpened to conduct a businesslike meeting.

Arrangement of meeting room

There are significant psychological influences arising from the form of a meeting room.

Some are obvious. For example, the physical conditions of appropriate heat, light and ventilation influence the members and the way they perform.

Less obviously, the shape and size of the table are influential. A long narrow table makes it difficult for members to have good eye-contact with one another and may prevent the chairman from being able to see all his colleagues.

The long square table, on the other hand, implies either that the chairman sits alone and apparently aloof in the middle of one long side, or that he is flanked by others. This is all very well if the others are officials, such as the secretary of the meeting. It is all very well too if the organisation to which they belong is highly democratic. But if the organisation is authoritarian, if the chairman is thought to be powerful and power-wielding, then the pattern of subordinates sitting beside him can be psychologically disturbing.

The shape should thus normally be rectangular, or possibly circular.

The size too is important. There is a natural spacing between people, a spacing which people find psychologically comfortable. If they are spaced too close together, elbow to elbow, it is not only physically uncomfortable; it imposes a degree of heat and tension into the meeting.

On the other hand, space them too far apart, and the members become remote, the discussion academic. They make long, discursive statements, interacting relatively little with one another.

The ideal spacing between chairs is about one foot or

slightly less. The chairman should ensure that chairs will be set at such spaces and that surplus chairs are removed so that members do not scatter themselves excessively around an oversize table.

After the meeting

It is a responsibility of the chairman to ensure that minutes are distributed, though the responsibility is normally delegated to the secretary.

For business meetings it is important to ensure that they are minutes which reflect accurately the chairman's understanding of what was said and decided. It is not acceptable to 'doctor' minutes of business meetings.

There are differences of view about how much detail of discussion should be minuted, but not about the minimum: the minutes should at least summarise decisions reached and the responsibility for taking action. Some chairmen find it helpful to specify responsibility alongside each minute, by insertion of the responsible executive's initials in an action column.

Finally, it is normally the chairman's responsibility to check that such action is being and has been taken after the meeting.

Summary

Outside the meeting, the chairman should:

1. Ensure that the routines are effectively conducted. Dates and venues chosen and notified, members consulted; agenda formulated and distributed.

2. Plan the discussion. Use the familiar three-phase preparation process:

   Random thoughts in phase 1;

   Analysis and structuring in phase 2;

   Reduction to stark form in phase 3.

3. Prepare for each item of the agenda, a card summarising:

   Purpose

Plan

Limits

Sponsor

4. Ensure the meeting room is well set, both physically and psychologically.

5. Follow up after the meeting, checking minutes and checking action.

15

Chairmanship
(2) Control of progress

The effectiveness of a meeting relies heavily on the effectiveness of a chairman. His effectiveness depends on this preparation, on the mechanics of his control of progress, and on the sensitivity of his control of members.

This chapter is concerned with the second of those ingredients, the mechanics of his control of progress.

The sequence of this chapter is:

1. The chairman's objectives and the role which he takes
2. The mechanics of control in the crucial opening moments
3. Sustaining control through the meeting
4. Concluding the meeting

The chairman's objectives and role

The chairman's strategy in handling a meeting depends on the objectives which he sets himself and on the role which he then takes. There are various ways in which he can 'play it', and his objective is critical.

For business meetings, that objective should be:

To reach unanimous assent in minimum time

Unanimous because business is about people getting things done, people working together. People do not work in harmony if differences between them are constantly being shown up. The chairman's job is to get them together, not to set them apart.

'*Assent*' is a key word in this definition of purpose—'assent', not 'agreement'. In a business meeting with reasonable, competent people representing different interests, it is right and proper that there should be differences of view. It is right and proper that this should be expressed, and it is wrong for a chairman to expect that unanimously all will *agree* when a course of action is against the interests of some. But reasonable people will accept that the majority view of the meeting is in a particular direction. The chairman should therefore aim to identify the feeling of the majority and should then get the minority to accept that this is the majority view.

Not:

'Will you agree to that, Harry?'

But:

'Harry, will you accept that the majority view is ... ?'

And if Harry now says 'No' then Harry is not being reasonable and the other members of the meeting will soon squash him.

Note the difference of atmosphere which the chairman can create by using this strategy of 'unanimous assent', in comparison with taking a vote and so emphasising the split within a work group.

In minimum time—for obvious reasons.

The chairman's strategy, therefore, should be to achieve unanimous assent in minimum time.

What sort of role should he be taking? Should he be the decisive authority? Should he be the key supplier of information? Should he be the key source for creative ideas?

For effective meetings, the answer to all three questions is a firm 'No'.

It takes intense concentration and intense skill to handle the procedures of a meeting and to handle the members effectively. Anybody who tries to be a key figure in the content of the meeting at the same time, is overloaded. Not 'may be' overloaded—he is overloaded. It is impossible to do so much.

An inevitable consequence is that the meeting is inefficiently conducted, the members frustrated, and time wasted.

Opening procedure

The character of a meeting is very quickly established. Within a couple of minutes of the members sitting down and starting, the pace and mood have been set, for better or for worse.

The chairman is critical in this establishment of character. In this chapter we are concerned with processes which he can follow mechanically whenever he is opening a meeting, without having to think about them. Repeated use of these processes enables him to concentrate on the personal element of his control of members.

At the outset of the meeting, as a drill, he should welcome the members and set a businesslike tone for the meeting. He should help them to focus together, reminding them of:

1. The purpose for which the meeting is called
2. The agenda
3. The expected length

Setting a target time has a strong influence on the time which will actually be taken. If the chairman proposes a length which is difficult but possible, then immediately he has members focusing on the speed with which they will need to move through the meeting.

He should both remind them of the purpose, agenda and time, and get their assent. In getting that assent, making a point of it, he establishes this as a meeting which is already unanimous—a meeting whose character should be one of continuing unanimity.

Occasionally, some member will want a modification, possibly to the form of the agenda or to the sequence suggested. Members should be encouraged to make such points at this time. It both helps to establish the momentum, and gives a chance for the chairman to demonstrate his flexibility and his search for unanimous assent. Listen to the member, check whether he has support, aim for unanimous assent to the agenda.

The chairman then leads through the agenda item by item. At the outset of each item, he should suggest the shape of the discussion as he has already planned it. Listing the purpose, why the item is on the agenda, the headings he suggests for discussion, and the limits to anticipate and impede red herrings.

He should get assent to his headings, not being afraid to modify them if members have conflicting views, and assent to his limits.

He thus starts the discussion on each item with a purposeful framework established, one which will also give him a firm basis for his control of the progress on that item.

Progress

In business, where it is important to have the members of a meeting pulling together, it is also important that their ideas and their thinking should interact with one another. The chairman should thus seek to have a progressive discussion, rather than a separatist discussion.

In a progressive discussion the contribution of each member follows on and picks up from the previous contributions of his colleagues.

In a separatist discussion, on the other hand, each member makes his contribution in isolation from others or even in competition with them. There is no interaction between successive contributions, and so a highly formal pattern tends to stifle creative interaction between members.

I find this particularly strongly in some overseas countries where there is a ritual of 'giving the word' to each member in turn—each then making an independent, possibly prepared statement.

As the discussion develops, the chairman should be seeking to maximise the total contribution by getting the best from each member; and constantly he should be searching to find the direction in which the weight of members' views is pointing.

If, for example, he has five members and he quickly senses that two want to follow course A and a third (member 3) to follow course B, he should quickly focus towards the remaining two members. He turns to member 4—

> 'Well Arthur, it seems to me that Harry and Jack want to go along course A. Is that how you see it?'

If yes, then the chairman simply has the job of getting first the other neutral, member 5, and then the dissenting member 3 to assent to the fact that course A is the majority view.

If, on the other hand, member 4 (Arthur) wants course B—then member 5's view is the critical one which the chairman must now bring in. And so, very quickly, he can sense where the balance of opinion is pointing, and lead the meeting towards unanimous assent.

This development is of course helped by having broken the discussion down into three or four sub-topics in his planning. Experienced users of this approach can assent to a sub-topic in a couple of minutes, four sub-topics in eight minutes; whereas a rambling discussion on the whole of the topic, inefficiently structured, could well last an hour.

The chairman also has the tool of the limits which he prepared. He is thus able to prevent the discussion going down some of the blind alleys which it might otherwise take. Having prepared a sharp picture of headings and limits, he is able purposefully to keep the discussion on the rails.

As the meeting progresses, he should keep the members conscious of the progress they are making. He should feed back to them the issues they have so far agreed, helping to underline the sense of unanimity. He should compare this *agreed* progress with the *agreed* plans and comment on it in the context of the *agreed* time.

Through the middle of the meeting, then, the chairman should be seeking for a progressive discussion. He should be constantly searching for the majority view, seeking assent to the fact that it is the majority view, emphasising progress and unanimity.

Ending

At the end of each item, the chairman should:

1. Summarise what has been agreed.
2. Ensure that action has been defined, what is to be done, by whom, by when.
3. Check that the secretary has a record.

At the end of the meeting he should:

1. Remind members of the purpose of the meeting.
2. Remind them of the time they planned to end the meeting.

3. Congratulate and thank them for having achieved the objective within the time limit.

Summary

In controlling the procedures of the meeting, the chairman's objectives should be 'unanimous assent in minimum time'.

Ideally his role should be that of lubricant to the discussion. He should not attempt a heavy responsibility for the content of the meeting as well as its processes.

He should open the meeting purposefully, and equally purposefully open each successive item. Key words here include

Purpose

Plan

Time

Limits

Assent

Through the middle of the meeting he should encourage a progressive discussion, looking for early and unanimous assent on each sub-topic and thus on each topic.

He should constantly be summarising progress in comparison with agreed plans, constantly emphasising the spirit of assent and agreement.

At the end, having summarised and defined action on each item, he should identify and comment on what the work group has achieved together.

16

Chairmanship
(3) Control of members

The chairman's objective should be to help the meeting to reach unanimous assent in minimum time. For this he needs the discipline discussed in the previous chapter.

Possibly even more important, he also needs the sensitivity and the skill to help the members of the meeting to express themselves and to reach that assent speedily.

This chapter is concerned with that control of members. The sequence is under the headings:

1. The character of a business meeting
2. Opening the meeting
3. The central phase of the meeting
4. Degree of formality

The character of a business meeting

Engineers soon become accustomed to attending meetings. Sometimes they find those meetings slow, frustrating, repetitive. Other times they attend meetings which are positively paced, creative and decisive.

It is only the skilled chairman who is able to move the meeting in such a positive creative and decisive way. In doing so his skill establishes four characteristics.

1. First, the tempo of the meeting. The pace needs to be businesslike. It must not stutter along lethargically with gaps between successive comments, or with slow and protracted statements from the members.

On the other hand, it must not be rushed. The desirable pace is the measured progress of a businesslike meeting.

2. The mood must be purposeful. The members must see the direction in which they are heading and must have a consistent sense of progress in that direction.

3. The meeting must be amicable. Members must feel free to voice their views, yet do so in a co-operative and cordial manner.

4. The mood must be co-operative. Members must be taken forward harmoniously despite likely differences of view.

Here then are four characteristics which the chairman should aim for in the character of his meetings: *business like, purposeful, amicable, co-operative.*

How does he achieve these characteristics? The potential for the meeting is very quickly established—within seconds, or at the most a couple of minutes after starting. Thereafter the chairman needs to exploit the potential through the remainder of the meeting. Let us look at the two phases independently.

Opening the meeting

The chairman's entry and his opening of the meeting has a crucial influence.

He comes in when members have been assembling, coming from their different tasks, each with his own worries and concerns. There are probably small groups amongst the members, talking, whether animatedly or desultorily. Each small grouping has its own interests, not yet fused or focused on the topic of the meeting.

The chairman's bearing as he enters the room is seen and is interpreted subconsciously. If the chairman's entry is hesitant, slouching, slow moving, his first impact is seen to be uncertain. He has already created a prospect of lethargy.

His entry therefore should be upright, confident, well paced. Not of course at that opposite extreme of overbearing and rushed which would precipitate a hectic and argumentative mood, but with bearing and pace reflecting the business-like qualities he wants to induce.

His timing is critically important. As he sits down there is a moment when most members are turning their attention towards him while others are still gathering their papers, and yet others are carrying on a private conversation. He should not try to start the meeting while these distractions linger on.

He must get the attention of the members harnessed. Not by trying to talk above their rustlings; not by asking for their attention; and especially not by asking the chatterers to be quiet. Any of those actions are felt by the members to be competitive with their interests and so those openings lead to a sense of discord between chairman and members.

The attention is best gathered by very careful timing. The chairman sits alertly, smiles at those who are giving him their attention and looks at the inattentive. Shortly, the attentive also look at the inattentive. The inattentive then become aware of a pressure which is felt to be coming from their colleagues. They look up. With a quick smile to them and a further glance all round, the chairman starts his opening comment.

There is a great knack in this timing. It normally takes five or ten seconds for the chairman to gain all-round attention in this way (though to him, waiting for that attention, it may feel like a lifetime).

There is then the knack of timing his opening comment to the split second.

What opening comment? It should be one which brings members immediately to a consciousness of the purposeful character of the meeting. Quickly, the chairman should make them aware of the direction the meeting will take and get their assent and approval.

He should have the headlines for this on a postcard in front of him, and his tone should reflect the positive and purposeful mood which he is establishing.

Using examples developed in Chapter 14:

> 'Good morning, gentlemen. We are here today to discuss the tender for Southport Sewage Works. You will remember that at our last meeting we agreed we would prepare such a tender. We are here today to agree on

how that tender will be prepared. I suggest that we will need to discuss:

(a) How keenly we should tender

(b) Availability of materials, plant and labour

(c) Whom we should suggest for staffing

(d) What we need to get done by when

Is that all right, gentlemen—will those headings enable you to make all the points you want? Harry, is that OK for you? Jack?'

(The way in which the chairman is using words here is important—but let us work through the rest of his opening before commenting on his choice of words.)

Good. We are agreed on the sequence of the discussion then:

Keenness to tender

Availability of resources

Staffing

Schedule

There are some limits which I think we ought to set, gentlemen. For example, we do not want to talk about the troubles we have been having with Northend Sewage or, at this meeting, about general problems with sub-contractors, do we?'

(Get assent).

My impression is that we should be able to do all this and come to a decision within 45 minutes. Is that reasonable for you, gentlemen?'

(Get assent).

'Right, gentlemen, so the first item we agreed to discuss is how keenly we should tender. Who would like to start us on this one? Harry, I think you have been doing some work on this one, haven't you? Can you tell us ... '

The sequence of points from the chairman here has established:

The purpose of the meeting

The direction of the meeting

The limits

The pace

He has on the way had unanimous assent to this procedure.

If there were any other views from members, additional points which they wanted to raise, for example, he should have tackled them flexibly, either extending his plan or incorporating their points within one of his headings (and making a note of the variation on his postcard).

In addition to setting out the procedure, getting and stressing unanimous assent, the chairman has also been taking steps to establish the momentum of the meeting.

His choice of words and the sequence of his words have been carefully chosen.

For example, when discussing the plan, the words suggested were 'Is that all right, gentlemen—will that enable you to make the points you want to?'

He asked the question 'Is that all right, gentlemen?'—reasonably sure that he would be surrounded by blank faces while they were still gathering their thoughts and recognising that some response was needed. He added the next question: 'Will that enable you to make the points you want to?' The second question is logically redundant—it had to all intents and purposes been asked by the first question. But the redundant question, taking another two or three seconds, gives the members time to harness their thinking and possibly to offer a response. Silence is a deadener of momentum, and the repetitive question is an insurance against it.

It is important for the chairman now to have a response. The meeting must be kept moving, the members must immediately become involved. But he cannot afford a gap, a break in tempo, a dreadful hiatus.

So, if he has not immediately got a response from the general body of the meeting, he quickly turns to a member who can be expected readily to voice a view.

    'Harry, is that OK for you?'

not

    'Is that OK for you, Harry?'

Harry, like everybody else, has been gradually attuning to the evolution of a meeting in which no ordinary member has yet spoken. If the question is posed with the suffix 'Harry', then he suddenly finds himself pitched into the middle. He needs a second or two before he responds; the hiatus has happened. The smooth flowing tempo has not been properly established. What is worse, the chairman is sensed to have hassled Harry.

So, instead, the chairman has carefully chosen the person he will first invite to speak, has spoken his name first, and has posed the question so that Harry will be in a position quickly to respond.

Just in case Harry does not immediately answer—the chairman should take further steps to prevent the development of a hiatus. He should ramble on for however many seconds are necessary, with his face, hands and tone of voice all inviting Harry's comment.

> Chairman: 'Harry, is that OK for you? Will it enable you to make all the points you want? Can you manage ... '
>
> Harry: 'Yes, that is all right for me.'
>
> Chairman: 'Jack, is that ... ?'
>
> Jack: 'Yes, fine.'

The mood is thus quickly established.

Jack now is quick to come in to stop the chairman's rambling. The meeting is beginning to move forward purposefully.

And so, through the procedures of establishing purpose—plan—limits—time, the chairman leads the meeting towards substantive discussion.

Again, the way in which he solicits the opening substantive comment must help to develop the mood and the momentum for which he has been working so hard.

> 'Right, gentlemen. Let us then start as we have agreed by discussing how keenly we should tender. Who would like to start on this?'

(Very brief pause. Take advantage if some member really will help get the meeting moving. But more likely the chairman will need to take one more step to give it impetus. So ...)

'Arthur, you have been involved with a lot of our tenders. I wonder what your experience is with tenders such as this. What is ... ?'

Ramble on momentarily, an interrogative in the voice, not too loud, eyes fixed on Arthur. He knows it's up to him and he'll quickly supersede the rambling.

So the momentum is established. The meeting is moving purposefully forward.

In the opening phase then, the chairman's behaviour is crucial. To summarise:

1. He needs to have a drill, a sequence which he can regularly follow while his conscious energy is elsewhere. The sequence, which should be getting fixed in the reader's mind, is:

   Purpose

   Plan

   Limits

   Time

2. His preparation should have enabled him to have the outline of these points simply and sharply stated on a postcard in front of him.

3. His conscious energy should be used to influence the mood of the meeting.

4. Key factors in that influence are:

   His posture and pace as he enters

   His attitude and timing as he sits and starts the meeting

   His establishment of assent amongst the members

   His subtle establishment of the momentum, the tempo, of the meeting

The central phase of the meeting

This section will be in four parts:

1. Impartiality

139

2. Manner of control of members

3. Types of member

4. Timing and assent

IMPARTIALITY

The chairman will often have a strong interest in the result of a discussion. He will have personal views about the important factors in the discussion and about the best outcome. He will of course be very tempted to offer these views to guide the members towards their decisions.

He should not.

He should be seen to be impartial.

The damage done by a chairman seen to be partial is of two forms.

First, it is natural that the members will react to his views and be over-influenced by them: either by suppressing their own views, in the face of a tough dictatorial figure, or by reacting and counter-attacking. However reasonable a person may be, when in control of a meeting he is seen to have a trapping of power. Unless he sensitively shows his impartiality, that trapping will attract emotional reactions.

Second, if he gets embroiled in the content of the discussion, the chairman will be challenged and find his energy absorbed in the content. He will not have the energy or the capacity to control the progress of the meeting as well as to involve himself in that content battle. It will be a bad meeting.

He must be seen to be impartial. This does not of course mean that he must be impartial: he may well have strong feelings about the matter, but he is not obliged to disclose them. Nevertheless, he can often powerfully influence the outcome by subtle use of timing and technique.

> If he has a meeting of five members, all wanting to go the opposite way to his, there is no sense in fighting. Accept quickly and do not waste time.

> If all five are of his view—quickly congratulate them, tell them how wise they are, emphasise unanimity and move on to the next item.

140

But if the meeting is split, two wanting to go this way and two wanting to go that way and the fifth havering—then wait for a powerful point to be made by the side he favours. Then turn quickly to the fifth member—'Well, Fred, that sounds like a powerful point in that direction. What do you think?'

And as, you hope, Fred reinforces, turn to the others and ask them whether they accept that this is the majority view.

The chairman must be seen to be impartial.

MANNER OF CONTROL

He must also be seen to be friendly.

His manner of control should be relaxed and cordial. A different chairman, one who is seen to be domineering and tight in his control, raises the emotional level of the meeting. His members' reactions against him and his attitudes prevent the development of a creative spirit within the meeting.

The chairman should always wear kid gloves; never the mailed fist.

Once he has the meeting running smoothly, a great deal of his control can be non-verbal. His posture should be consistently alert, upright, forward to the table. Never pensive, never slouched; reclining very occasionally and very deliberately, for reasons shortly to be discussed.

His eyes must focus attention towards the speaker. He must be rapt in that attention and his energy must permeate to the other members so that they too are rapt. But while focused almost entirely on the present speaker, his attention must also be alert for other people wanting to make a point.

He should use facial gestures, particularly, to encourage and support whoever is speaking. The friendly nod, indicating understanding of the point being made (not necessarily acceptance); the supportive smile; the inquisitive raising of the eyebrows; the range of facial expression which can reflect and support the speaker.

Posture, eye-contact and hand gesture should both support the speaker and lubricate the discussion. As he wants to bring in another member, he avoids excessive use of his voice. He

identifies the next speaker with an eye-contact, a hand movement and a change of body position to face in that speaker's
direction. His non-verbal control of the meeting can then
keep the focus of all members' interests pointing towards each
successive speaker. He can keep a meeting moving with strong
momentum, without himself having to utter a word.

Through this central phase, it is important that all
members become actively involved. Recommended procedure
here is for the chairman to have a blank sheet of paper handy.
At an early stage, to draw a rough layout of the meeting table
and of each position occupied. Then to put one mark against
each position as that member speaks.

It can be done quite casually and with barely discernible
loss of concentration on the current speaker. Quickly the
pattern becomes plain. The chairman has evidence of who—if
anybody—is out of the discussion and can take steps to bring
him in.

As usual, first identify him by name, then pose an easy-to-
answer question. And ramble until at last the chap breaks his
silence.

> 'Douglas, I wonder how you hear these arguments
> stacking up? Do you think that the issue about the
> water main is an important one? Is it the sort of thing
> that your people have come across very much? Would
> you ... ?'

The general principles then on which the chairman should
handle the centre part of the meeting are:

Impartial

With kid gloves

Heavy use of non-verbal communication

Bringing everybody in

TYPES OF MEMBER

How does he apply these principles in the face of some of the
difficult characters to be found around any meeting table?

There is the loquacious character, the chap who goes on
talking for ever.

But even he has to pause for breath. The moment he does
so, even in mid-sentence, the alert chairman has turned to a

member whom he knows to make pithy comments. In turning, the chairman's shoulder should have moved so that it is beyond the loquacious speaker—he should just be offering his back for the loquacious one to talk at while his (the chairman's) energy, concentration, eyes, are focused on the man who will make the pithy comment.

No need for the chairman to speak a word.

There is the other loquacious character, the one who makes far too many comments.

Preferably do not recognise him. Use non-verbal technique to switch attention amongst the other members of the meeting.

This loquacious one is often insensitive. He will bluster on, regardless. Then, in fairness to all the other members who want to speak and who need the chairman's protection, the loquacious one must be stilled.

> 'Just a second, Jack, before we go on to new ground, I'd like to know what Mary thinks about the previous point about . . . Mary, what are your views . . . ?'

There is the intransigent member, the man who is determined to go off on his own hobby horse.

The chairman must not alone squash this man. He must have him squashed by the meeting as a whole. He does so by referring to the agreed plan or the agreed limits, and then invoking the support of the membership.

> 'Yes, Fred, if I understand right, you are making the point that the presentation of Proposals should be in better printed bindings. I can understand the point, but we agreed that today we should discuss scope and responsibilities. We have agreed scope and we are now discussing responsibilities. Well, gentlemen, do you want me to add bindings to our list of headings to be discussed within these 45 minutes?'

The intransigent one might have been awkward at an earlier stage, while the plan was still being discussed. He might have pushed for his hobby horse to be made part of the discussion.

It might have been relevant. If so, the chairman should have been flexible, adjusting his plan accordingly.

Or the chairman might have considered it irrelevant. He then has two options. One is to check whether the majority would want to give time to that item, then get Fred to accept the majority view.

Alternatively, there is another device which is particularly useful for 'straw issues'—molehills which somebody is trying to make into mountains. It is undesirable to focus the meeting's concentration, energy, time on such straw issues, even if it is only to reject them. The device is for the chairman simply to accept the point and to put a note of it at the end of his plan for the discussion. My experience is that when, after 44 minutes' discussion has taken us purposefully through a difficult topic, I finally turn and say, 'Well now, before we break, there was another point you wanted to make, wasn't there Fred? Does it still seem important?'—Fred has forgotten. The way I voice the question does not help him to remember. His colleagues are all packing up their papers. We still get away at 45 minutes.

There are the privateers: two chaps, sitting half way down the left-hand side in a meeting of a dozen people. They start their private conversation *sotto voce*, while Harry is speaking, and distract everybody else. Quietly, firmly, with kid gloves, they must be put in place. Immediately.

Immediately, a quick glance towards them. This just may be sufficient to regain their concentration, to bring them back to responsible behaviour. Most probably, further steps will be needed.

First, stop Harry. Look back to him. Hold up a hand to him. If necessary, a quiet word to stop him. Cordial, friendly, kid gloves.

Now, silently, get the meeting to impose discipline on the privateers. Posture and body position should be directed towards Harry, so sustaining his paramountcy, hand held up steadyingly towards Harry, but eyes focused on the dissident duo. The effect is that nine other pairs of eyes swivel towards them. Nine other members, including Harry, resent the duo. That is a powerful force acting on them, a force they can feel. Guiltily they look up, see the chairman and colleagues staring at them. Discipline is re-asserted.

Now the chairman reinforces his control and the friendly nature of the meeting. He smiles at them, and then switches attention firmly back to Harry. Body position towards Harry, face and eyes towards Harry, feeding back the discussion for the benefit both of Harry and other members while quietly and implicitly rebuking the duo.

'Sorry, Harry, we were talking about the inclusion of stop taps, and you were making a point, I think?'

Hand open towards Harry for him to continue.

There is mishmash. Two or three separate pairs of people start whispering to one another, making it quite impossible for the meeting to continue. Under competent control, it is rarely a phenomenon. Nevertheless it can happen.

The previous unobtrusive handling might still be made to work. But if not: the chairman must regain control, but as always, with a kid glove.

First, stop Harry. He will be well aware of the hubbub, wanting the chairman's protection. It is not reasonable for him to be expected to continue against such hubbub. Stop him with a gesture and a smile of resignation.

Now lean back and leave the hubbub to go its own way. Quickly people will realise that something is wrong. They will glance towards the chairman and see him, extraordinarily, leaning back in his chair.

Then as the hubbub dies, timing is again critical. He comes forward, smiles benignly and says:

'Well, gentlemen, it will be much more effective if we make our contributions through the Chair, won't it?'

And as their understanding and sympathy grow, he brings the meeting purposefully back to its agreed direction, and then re-establishes Harry for his contribution.

'You will remember that we have so far agreed that we are very keen to get this contract and we agreed to go on to discuss availability of resources, then key staffing and the schedule. Harry, you were making a point about the availability of labour. Would you care to remind our colleagues ... '

These are the techniques for handling the difficult members who are likely to be a part of any meeting—handling them always with kid gloves.

TIMING AND ASSENT

Timing—short-term timing, getting the split second right for a gesture or a remark of the chairman—has been emphasised as an important element of his knack of control.

Timing is important in the other sense of the meeting as a whole. The chairman should note on his postcard the starting and target ending times for each item. Alongside, he should have his watch. With the two in front of him, he is well equipped to control the pace at which the meeting develops. In fact he will find that the subconscious takes care of this for him. He can fix his conscious energy on the way in which he is handling the members, and will find—occasionally to his surprise—that the schedule is fulfilled.

It is important throughout the central phases for him to be seeking for assent, emphasising the mood of assent, emphasising the assent already achieved.

Degree of formality

Surname or Christian name?

It is a question of what is normal for the community. Many engineering organisations with proud and respected images have a tradition of behaving formally. For them, the meeting should reflect the formality. Title and surname should be used.

But the trend in modern business is towards more informality and towards the use of Christian names. It makes for an easier, more informal meeting.

Formal processes, such as proposing and seconding resolutions and amendments to resolutions, should not be necessary in most normal business meetings.

Should all contributions be through the Chair?

Indeed they should. Even in the most informal of meetings, it is imperative that the chairman be in control. It may be a relaxed control. He may be trying to lubricate rather than dominate the discussion. He must nevertheless be in control.

Members should occasionally support him, even in informal meetings, by using the word 'Chairman'.

Summary

In his control of the members, the character of meeting for which the chairman should aim is:

Businesslike

Purposeful

Amicable

Co-operative

The chairman's opening of the meeting has a crucial influence. There are lots of skills which he needs to use: timing, establishing assent and unanimity. Establishing unified views of the purpose, plan and pace of the meeting. Careful choice of words to establish momentum.

Throughout the meeting the chairman should:

1. Be seen to be impartial

2. Be seen to exercise his control lightly

3. Sustain a co-operative atmosphere

4. Be alive to the needs and possible contributions of different types of member

5. Be meticulous in his timing—both split-second timing and keeping the duration of the meeting within bounds

17

Behaviour in meetings
(1) Case-presentation

Engineers attend many meetings at which they must present themselves and represent their organisations effectively.

This chapter is concerned with the way in which they project themselves when they are 'in the spotlight' presenting a case or a report during a meeting. The subsequent chapter will deal with behaviour at other times during meetings.

There is an enormous difference between the behaviour appropriate to two different sorts of case presenter. They are:

1. The *expert*. The person expected to have special knowledge and experience and whose task it is dispassionately to advise people lacking his own expertise.

2. The *advocate*. The person expected to make firm proposals and to press for their adoption.

These two different types of case presentation demand both differences of technique and differences of attitude. They will be considered in separate sections:

1. The methods of the expert

2. The methods of the advocate

The methods of the expert

The expert's task is to assemble and offer evidence, and to give advice on interpreting the weight of the evidence.

The reputation he needs is that of being authoritative, experienced, clear-minded. He needs to appeal to that part of the brain which handles rational thought.

It is likely that his statement will be published, and it must be capable of standing up to the scrutiny of critics, whether public or private.

It must therefore be full, precise and accurate.

There is a duty on the other party—be he listener or reader—to study the statement in a way which is impossible with a once-heard talk or lecture.

The expert's choice of material therefore needs to be an exhaustive selection of the evidence and an interpretation of that evidence. The rule is, whether or not his version is to be printed, that there is yet merit in his writing it in full. The approach for these purposes should follow the methods suggested in Part 2 of this book, 'effective writing'.

We have previously seen that effective communication depends on a message being put in terms which are sharp and simple, and that this general rule can be relaxed a little for the written word, which can be studied and re-studied. At first sight, therefore, it seems that the expert does not need to simplify for presentation of evidence.

He is, however, likely to be questioned and cross-examined on his evidence, possibly by people experienced and skilled in the art of demolishing evidence. Under these circumstances, he needs to have sharpened his own thinking. If he goes to a meeting confident that he knows the whole situation, yet not having sharpened his thinking, then he will be caught by sharp questions. He will find himself trying to marshal responses, taking account of many shades of subtlety, spread over many variables. All too easily, his responses can be heard to be rambling and diffuse.

To avoid that image, he must sharpen his thinking. He should use a normal preparation process leading to his evidence being marshalled under a few clear sharp headlines.

In choosing his material then, the expert must give exhaustive treatment in a written form, and he must also simplify his thinking so that he can later cope with critical cross-examination.

What about when he comes to deliver the case during a meeting?

The image he is seeking is that of the authoritative, the experienced, the clear-minded.

His natural authority can be reinforced by the way in which he used some symbols of power. For example, if possible, he should stand rather than sit while giving evidence. He should make use of slides, graphs, charts, visual aids. They are powerful aids and they can add to his stature, but they must of course be elegant. An untidy aid would give an impression that the speaker was casual—not expert!

His credibility is, as ever, critically influenced by the first impressions which he creates.

Often his opening remarks have to be a self-introduction. Many engineers are reticent, even humble people and it is all too easy for them to understate their competence.

What the listeners need to know is:

Who he is

Where he comes from

What his qualifications are

What experience he has had relevant to the case

Not

'My name is Jones and I am a mechanical engineer.'

But

'My name is Rodney Hugh Jones, of Arbuthnot, Jones and Slater, Consultant Engineers. I hold a Master's Degree from the University of Cambridge and am a Fellow of the Institution of Mechanical Engineers. I am a Senior Partner in the Partnership and have 17 years' experience in handling the problems of . . . '

He needs to make the statement confidently, clearly, positively toned, positively paced.

Confidently. Erect posture, head up, shoulders erect (not aggressively up or forward—certainly not drooping or dejected).

Clearly. The words articulated so that they are heard. Do take three deep breaths on your way to the stand and start speaking with a reasonable lungful of air.

Positively intoned. The statement should be intoned on rising levels, both of pitch and volume.

Not: **'MY NAME** Is Rodney arthur jones. **I HAVE A** Master's Degree from cambridge.'

But: 'my name Is Rodney Arthur **JONES**. I have a master's Degree From **CAMBRIDGE**.'

Positively paced: The pacing should again be businesslike—not slow and stumbling nor rushed and apprehensive. There is merit in making heavy use of pauses. They can be used to give an impression of deep thought, weighty consideration of the weighty matters in hand.

These first impressions, arising both from what is said and from how it is said, are so important that the expert does well to rehearse them with a critic before he goes near the meeting.

The manner of the expert's presentation of the body of his evidence should be formal. There is of course some need for audience contact, but the expert should not be seeking the depth of personal bonding which is appropriate for the normal public speaker. His search is for credibility at a purely rational level.

Excellent examples of the distinction between formal and informal were to be seen during the time of the Falklands crisis. The professional spokesman from the Ministry of Defence adopted an attitude and a tone massively more formal than his counterparts in Argentina. He was seen to be the dispassionate expert, not the impassioned partisan.

There is room for argument about whether he overdid the formality. There is no room for argument that a degree of formality gives much more the impression of professional competence than a more emotive presentation.

The professional spokesman nevertheless was positive in his use of eye-contact, of gesture and of voice. His occasional use of a visual aid—the maps—greatly helped to concentrate the attention and the interest of his listeners.

The expert's preparation therefore should cover all the facts, possibly submitting them in writing and supporting with an oral statement. The presentation needs to be formal, positive, confident, authoritative.

The methods of the advocate

The advocate is in a different position.

It is not his job to review all the evidence, dispassionately. His task is to propose a course of action, to sway opinion in favour of that course of action.

The credibility of his case therefore depends on his ability to generate people's enthusiasm for it. The advocate thus needs to be quite different from the expert in his choice of material, and in his presentation of that material.

He should not be cornered into presenting a case unless it is one about which he feels conviction and enthusiasm.

Pity the poor engineer asked to advise, as an expert, on the routeing of a road, making his recommendations, then later being required publicly to present the case for a different route. He first assembles all the facts; he then analyses them and on the balance of all the evidence he recommends (say) route A. The decision-makers, weighing some of the non-technical factors differently, may opt instead for route B.

If the engineer then is asked to present the case for route B at a public meeting, there will inevitably be trouble. First, having previously and carefully prepared the ground as an expert, he may fail to make the different sort of preparation needed by the advocate. Second, and more important, the proposition of route B is bound to attract some hostile questions and comments. There are bound to be some who will suggest that route A would have been better, to throw in arguments which basically he supports. No matter how he tries, under such questioning, his lack of conviction is bound to show through. Both he and the proposition will suffer.

As a basic rule, try to avoid getting drawn into presenting a case unless it has your confidence and commitment.

His process of preparation should follow the normal pattern. First, the A4 stage of jotting random ideas; second, the A5 of thinking what will influence the listener and analysing under a limited number of headings, leading to the final 'prompt' of a postcard with a few words printed large.

His choice of material at the A5 is particularly significant. He will be able to see a dozen good reasons for following the course of action. He will naturally be tempted to project all dozen to the meeting.

He should resist the temptation.

Consider what will happen, supposing for illustration that there are half-a-dozen other people at the meeting. The topic is by definition controversial. It can be expected that the members will naturally fall into three groups: those who are *for* the case (call them Fred and Frank), those *against* (Alan and Arthur), and those who are *not committed* (Noel and Norman).

The advocate's success or failure will depend on the way in which Norman and Noel are swayed.

If he outlines his dozen reasons in favour of the case, an antagonist will unerringly seize on the least significant. He will attack it and continue to attack it. Being the least of his reasons, the advocate will defend it less strongly than he would a more important issue. He may even think it does not matter too much if he concedes point 12—there are, after all, another 11 good points. But the effect on Norman and Noel is that they see a prop knocked out from under the case.

And then, equally unerringly, the other antagonist weighs in and starts to attack point 11. Norman and Noel have already seen the advocate under pressure; they see him under further pressure; there is no need to go much further. The two critical members of the meeting have been swayed to doubt.

Consider what happens on the other hand if the advocate chooses only the four strongest points in favour of his case.

Strongly, he advocates them. When he stops, somebody—it might be Noel or Norman, more likely it will be Fred or Frank—will express his surprise that there has been no mention of point 5. The advocate, while accepting that 5 is a ground in favour, resolutely protests that he has given the four strongest reasons.

Then somebody comes up with point 6 in favour, and Noel and Norman feel increasingly that there is merit in the case.

Choice of material for the advocate should therefore be restricted to those points which most strongly favour his case. Resist the temptation to include the less strong, less defensible.

The case presenter also has the opportunity, while presenting his case, to minimise the strength of the opposition. He can anticipate one of their stronger points and set it in a context in which its significance is diminished.

'We have looked hard for a method which would avoid having to demolish three shops, but the only practical alternative would mean demolishing eight privately occupied houses.'

It is not only in his choice of material that the advocate should behave differently from the expert. His presentation also must be different.

He must try to sway opinion, and his presentation must appeal to parts of the brain which handle emotional impulses, as well as to the purely rational.

He must be seen to be enthusiastic, friendly, committed, practical.

As ever, first impressions are of crucial importance. When invited to present his case he must immediately convey his own feelings of commitment and conviction.

'Thank you, Chairman. There are in fact four reasons why we should build these works at point X. They are . . .'

The way in which the statement is made must reflect the character of the meeting.

In the conditions of most business meetings, the advocate is likely to be sitting at a table. His voice must be adjusted to those conditions and his tone must be intimate, friendly, persuasive.

He must have eye-contact to convey his enthusiasm and interest to every one of the members present. This is easy enough with people sitting opposite, difficult with people sitting on the same side of the table. It takes a very deliberate effort to make that contact with one's neighbour, and one may even have to shift position slightly to make eye-contact with the member who is further along and partly obscured.

However difficult, the eye-contact must be made.

The advocate's posture and gestures—manual and facial—must reinforce the friendly, intimate, yet confident and committed impact which he needs to make.

When he quotes figures, he should make those figures visible. If there need to be many, he is entitled to bring in and to hand round a paper showing the figures.

If the advocate needs relatively few figures, there is much

to be said for writing them then and there on a sheet which can be seen by the other members. This can reinforce the intimacy and the sense of 'invented here' which he should be trying to generate amongst the other members.

In his presentation then, the advocate should have his energy radiating and enthusing his colleagues around the table. He should be using his voice and his battery of non-verbal communications in support of his case.

Summary

If presenting cases to meetings, the engineer may be required to play the role of expert or of advocate.

These are different roles, needing different techniques. The expert's preparation should possibly veer towards a written statement to be accompanied by an oral over-view.

The expert needs to review the whole range of evidence and assess its significance. Probably also to draw conclusions. His presentation should be formal, confident, authoritative.

The advocate should avoid being burdened with advocacy unless he really believes in the case.

His preparation should concentrate on the few most positive factors.

His presentation should be friendly, committed, intimate.

18

Behaviour in meetings
(2) The ordinary member

This chapter is about behaviour which helps the ordinary member to shine during a meeting. It is under three headings:

1. Making contributions
2. Perceived presence
3. Helping progress

Making contributions

Any member attending a meeting may make much or little contribution to it. Indeed, he may make too much or too little.

The first step is to be well prepared. He should have studied the agenda and identified points at which he will want to contribute or will be expected to contribute.

Where the contribution is to be on a particularly important item, he should have prepared in depth. Even though not formally invited to make a presentation, he should yet prepare on some such lines as those suggested in Chapter 17, and should not hesitate to sound out and to seek support from colleagues who will be attending this meeting.

The structure of the contribution is, as ever, important. If he wants to say more than three or four sentences, his opening comments should alert the minds of colleagues for what is to follow; central parts of the message stated sharply and simply; and the whole briefly summarised.

The positive member of the meeting is of course regularly

involved in giving comments on matters which come up spontaneously. Such comments should always aim to be constructive and creative. Negative attacks, destructive criticism, earn reputations for negative thinking. The engineer should try positively to phrase his contributions, positively to point possibilities.

His frequency of contributions also influences a member's reputation. In any meeting there is always some individual who acquires the reputation for talking far too much. He is the one who is determined to speak on every point, forever taking the time of his colleagues. And then there is some other member who is seen so rarely to comment that he might as well not be present.

Somewhere between these extremes is the member who is ever ready to respond to a query or a hint from the chairman, without excessively pushing himself forward. He is the one whose views and whose presence will be sought.

Another fault is to ramble on at great length. Such rambling comes from a failure to sharpen and simplify one's thinking. One may perceive a situation as being complex, with a murky mass of material needing to be considered. The discipline needed here is that of clear thinking, breaking a topic down into its constituent variables and classifying those variables under a sufficiently small number of headings to be readily communicated.

Perceived presence

An aura surrounds the member attending a meeting.

Suppose he has his chair back six inches further than his colleagues, that he is sitting well back in his seat, with his arms folded, possibly even with his eyes closed.

He is seen to be remote from the ongoing business. His presence has a subtle influence on his colleagues, acting as a drag on the pattern and pace of the meeting. The aura which surrounds him is negative.

A positive aura can, however, surround a member who does not even speak during a meeting. It comes from the way in which he is seen and felt to be behaving.

Chair positioning is one ingredient. It is not simply a question of whether the chair is drawn up to the table or pushed back a little (respectively 'present' and 'distant' relationships to the meeting). It is also a question of the angle at which the chair is positioned. The member sitting at the end of one side of a table, for example, might turn the chair slightly so as more clearly to face the middle of the table. It is a positive act. It conveys a message of enthusiasm and commitment.

Posture should reinforce the impression of lively interest. Upright, giving the lungs every chance to get a full quota of oxygen. Hands visible—probably on the table—where they can readily make gestures. Yet seen to be relaxed—not raised or hung shoulders.

Gestures, especially facial gestures, are important even when not speaking. A positive aura is helped by nods to convey understanding of what somebody is saying, or by quizzical looks, or by smiles. Whatever the particular mood being conveyed, the face can express animation or inanimation.

As always, eye-contact is powerful. The ordinary member who makes a point of concentrating firmly on each successive speaker, making positive eye-contact with each, is seen and felt to have strong involvement and interest. He is seen to be positive.

Such are the ingredients of the aura which surrounds any member of a meeting. The purposeful member is constantly using this non-verbal communication to earn a positive aura.

Helping progress

Each member can also influence the progress of the meeting.

The momentum of the meeting is dominated by the chairman and his behaviour. He must have support and competence from members if that control is to produce a satisfying momentum.

The chairman will in particular be conscious of that member who can be relied upon to offer short, pithy, relevant comments on request. The competent chairman will make great use of that member. Working in harmony, they will get and keep the meeting moving.

The ordinary member can be a great help in other ways to the progress of the meeting.

It is the chairman's job to ensure progress according to

some plan and time scale, but in the hurly-burly of a meeting, with competent engineers (possibly laymen too) representing different viewpoints and different interests, possibly with a need to control some awkward customer, the chairman can readily become overloaded. He may lose track of progress. He may find it so difficult to involve the passive, to restrain the hyper-active, to rail off red herrings, that he loses awareness of the pace at which the meeting is moving.

In such a moment it is a tremendous help to the chairman and to the meeting if some member makes a 'process intervention': if he says something which is designed to help the process which the meeting is following. Examples of such statements which help the chairman to keep meetings in order, are:

1. 'I wonder, Mr Chairman, if we could have a summary of how far we have got on this point?'

2. 'Mr Chairman, I am concerned about the time. Is it possible for us to move on?'

3. 'Could I check what the secretary has recorded about this discussion so far?'

4. 'Mr Chairman, could we now take a decision on this issue?'

Another form of process intervention is that which helps the chairman to keep the meeting to the point. If one senses that the discussion is getting into a blind alley, it helps to offer such comment as:

'Mr Chairman, can we go back to the issue which you first raised—the issue of ... ?'

The ordinary member also can help the chairman to control the members. Examples include asking to hear the views of one who has been passive, or helping to subdue the whisperings of a dissident duo.

The character of a meeting is influenced too by the ability of some members occasionally to introduce a touch of humour. Not the repeated sarcasm of the acrimonious, nor the lengthy jokes of the buffoon, but the occasional brief spark of humour which relaxes.

Most engineers are conscious of the need to contribute to

the topic of a meeting. Relatively few show the competence to help the chairman with the process of a meeting, with control of its progress and of its members.

Summary

1. Behaviour in a meeting influences both the decisions in that meeting and the reputation of each member.

2. Contributions should be made positively, and not excessively in either frequency or duration.

3. An aura surrounds each member, for better or for worse, depending on the way his presence is perceived.

4. Each member can help the chairman to control progress and to control other members.

19

The effective secretary

The secretary's role is vital to the conduct of a committee.

His work can create for the members the possibility of smooth and effective effort. In his role, he can complement the work of an effective chairman or partly compensate for one who is less competent. His successes are never dramatically obvious, but his failures are.

The way in which he works must depend on the character of the chairman, and on his relationship with that chairman. They can share their responsibilities, and they need to agree how they will do so. For example, either or both together may take the initiative in ensuring that the agenda is formulated.

While the secretary may be appointed for a one-off meeting, the assumption in this chapter is of a secretary appointed for a succession of meetings of a particular committee.

His tasks are discussed in chronological sequence. What he does:

1. Before the meeting
2. During the meeting
3. After the meeting

Before the meeting

The secretary must first make sure that he understands the purpose, the authority and the limitation of the committee he is to serve. He must know the membership, and the frequency of meetings. If these fundamentals are not already clear, he must ensure that they become defined.

In advance of any meeting the secretary must distribute papers for that meeting. Key amongst the papers is the agenda. It should be discussed with and approved by the chairman and the discussion should cover any need for contributions to be prepared by members, either in writing for pre-circulation or for oral presentation.

An agenda normally includes:

> Title of meeting or committee
>
> Time, date and place of meeting
>
> Purpose of meeting
>
> Agenda items:
>
>> Apologies
>>
>> Minutes of last meeting to be considered
>>
>> Matters arising (if not covered elsewhere)
>
> Main items for discussion, in sequence, cross-referenced to supporting papers

Any other business

Details of next meeting

The chairman and secretary should agree which of them will solicit the necessary working papers, and the secretary will probably need to chivvy contributors. It is his task to ensure timing and distribution of the agenda and supporting papers, together with any necessary details of accommodation, maps, etc.

The secretary also has the responsibility for ensuring good physical arrangements: meeting room, tables and chairs, supply of paper, availability of refreshments, flip-charts or other audio-visual aids.

In summary, before the meeting, the secretary ensures that the meeting will start with sound administrative foundations.

During the meeting

The secretary's primary concern is towards ensuring that competent minutes can be distributed.

Even though the minutes may be circulated in summary form, the secretary is prudent to ensure that he has a full record. It both helps his later summarising and acts as insurance in case any member comes back later on a detail. A tape recording is a good insurance, even though it may simply be stored and never used.

As each item of the agenda is completed, the secretary should make sure that he knows exactly what has been agreed. It sounds easy. With a good chairman, it is easy; he summarises the discussion, defines the decision, and makes sure that the secretary has got a note of it. With a less competent chairman, decisions may be obscured and become clear only if the secretary asks for clarification and persists until he gets it.

There are of course committees in which decisions are deliberately obscured. It is a fine political tactic to let the meeting be vague and then go away and write minutes of what the 'in-crowd' would have liked to have decided. It is all very well for political meetings, but should not be normal behaviour in business meetings.

While his first priority is towards accurate minuting, the secretary has other interests during the meeting. He should not normally enter into discussions, but he has the responsibility for correcting factual inaccuracies, especially mis-quotations of previous practice or decisions within the committee.

He should try to anticipate any need for supplementary information during the meeting and should have spare copies of working papers available.

He can draw the chairman's attention to speakers trying to get into the discussion, or to people who should be able to contribute to a point but who have not made themselves heard.

He should be alive to progress. He should, if necessary, prompt the chairman, helping him to keep to the point and to keep on schedule.

The secretary also has welfare concerns. Heat and light, refreshment, summoning taxis, all are issues to which he may need to give some attention.

165

After the meeting

The minutes are the secretary's prime concern.

Current trends are towards summary minutes. Compared with full minutes reporting all of a discussion, summary minutes are less time-consuming both for the secretary and for the readers; and they are likely to be less suspect. Different people will hear and interpret points in a discussion in different ways, and detailed minutes are always suspected by somebody.

Summary minutes should include a record of decisions reached, actions needed, and responsibility for taking those actions. The secretary should if possible check with the chairman before having the minutes despatched, but it is his responsibility to ensure that they go out promptly. If they are not circulated within two or three days, his reputation will suffer.

The chairman and secretary should agree which of them will take the lead in checking that action is taken in line with the minutes. This could involve delicate follow-up and discussion with the members.

Between meetings, the secretary should be on the lookout for any material relevant to the work of the committee. He should if necessary arrange for it to be copied and circulated to the members.

Summary

The secretary has a crucial role in complementing or compensating for the chairman's actions.

Before the meeting he should:

Agree the agenda with the chairman

Have working papers prepared

Circulate agendas

Make the physical arrangements

During the meeting he should:

Keep the record

Help the chairman's control of progress

Look after the welfare of members

166

After the meeting he should:

Draft the minutes and check with the chairman
Circulate the minutes
Follow-up, checking action
Circulate any material helpful to members

20

Repetitive meetings

The committee which meets regularly takes on a life of its own.

That life has a consistent pattern. It is one in which great things may be accomplished at one time, nothing useful achieved at another.

What is this consistent pattern? What steps should engineers be taking to maximise effectiveness?

In this chapter we:

1. Outline the consistent pattern
2. Describe what happens in each phase
3. Suggest when and how to recognise danger signals

A consistent pattern

Repetitive meetings go through a series of four phases.

It may be that deliberate steps are taken to modify this normal development, but without such steps, the sequence is inevitable.

The successive phases are:

1. A creative phase
2. A regularising phase
3. Productive phase
4. Decay

These phases will be of varying lengths.

THE CREATIVE PHASE

In this opening phase, when a group is meeting for the first time or two, it is fresh and should be dynamic.

The need for the meeting is freshly perceived. Be it for decision-making or for co-ordination, it is bringing together a fresh group of people.

They may arrive knowing little of one another, needing to build their relationships and their ways of working together. Terms of reference may be formally established, but informally, the real purpose and methods of the meeting take on their own particular identity.

Tackling a new remit, the work group has the opportunity to make some sort of pioneering progress. This can be an exciting and enjoyable phase, building a sense of belonging amongst the individual members.

The duration of this creative phase is normally of the dimensions of a couple of meetings.

Members are still feeling their way towards what they need to talk about and how best to contribute. The agenda tends to be overcrowded, time to be tight.

THE REGULARISING PHASE

Having become established, the sequence of meetings soon begins to take on a regular form.

The form of the chairman's control becomes recognised. His relationships with the secretary and the secretary's contribution begin falling into a pattern. The frequency and duration of the meetings begin to conform to a pattern. Members begin to recognise one another's strengths and one another's peculiarities. They begin to form acquaintanceships and pairings.

Within another two or three meetings, what they do and how they do it has become a matter of routine.

THE PRODUCTIVE PHASE

Each successive meeting now conforms to a regular pattern. It has its own peculiar routine, developed for the particular task by the particular grouping of members under the particular leadership of their chairman.

It is a pattern in which they should be able to contribute effectively and expeditiously.

Through successive meetings, the way in which the members interact with one another takes on an increasingly stereotyped character. Recognisably, each begins to build an image. They may be seen to be trying to establish their own position or power; to be constantly allied with one another in predictable ways; to be constantly creative and purposeful— or constantly destructive and critical.

Much of the work of any committee is conducted outside the meeting room. Individuals collaborate, seek support, seek to diminish opposition. It is right and proper for people committed to the interest of their own department or project to take such steps to create a possibility of progress. But of course there is always the danger that this canvassing will become too dominant an activity. It becomes possible for some people to concentrate on the motions of canvassing while losing sight of the objectives of the committee.

As the meeting becomes routine in its form, so should it become possible for time to be used more effectively. If a particular discussion needed half a day in the infancy of a committee, it should be possible in its maturity to cover the same ground in a quarter of the time.

But gradually, the sequence of meetings drifts from being productive.

THE DECAY PHASE

The meetings go beyond being routine and become a ritual.

The original purpose of the meeting may have been satisfied. The surrounding business conditions will certainly have changed. The membership will have changed. In its early prestigious days it will have attracted eminent members. As it moves beyond the exciting into the routine phase, towards the ritual, so it becomes less attractive to key members, who begin to send their deputies along instead.

Nevertheless, in establishing itself, the committee established a capacity to get things done. It established some sort of authority, and that authority is seen to be sustained. Diminished it may be; die it won't.

The rituals of frequency of meeting are sustained. The duration established in the early stages also becomes a ritual. If it used to meet at 2.00 and go on till 5.00, there is a tendency for it still to meet at 2.00 and still to go on till 5.00, no matter that the justifiable time is now but a fraction of three hours.

Nevertheless, aided and abetted by the routine of 'Date of next meeting' always being the last point of the agenda, the sequence of meetings drags on.

Danger signals

This sequence of phases—creative, regularising, productive and decaying—is a consistent pattern for the life of a committee. Yet it is difficult to recognise when the time has come for change. What are the danger signals? They are of three forms.

One relates to the age of the particular committee or meeting. It moves from the regularising into the productive phase within at most half-a-dozen meetings. Thereafter there are many variables which will determine how long it goes on being productive: the nature of the task, the external pressures, the character of the chairman and of the members, their respective skills and relationships. But after a number of meetings, productivity is liable to be sinking. How many meetings will it take before this happens?

As a rough rule of thumb, it is worth asking whether the committee has served its purpose, or is in need of refreshment, after it has met a dozen times. The odds by then are that some at least of the symptoms of decay will be creeping in.

After a score of meetings, increasingly there is likely to be a loss of momentum or effectiveness.

The second form of danger signal is in the membership. If a meeting becomes enlarged, its tightness is soon dissipated.

Beyond a certain size—about eight people—it becomes difficult fully to involve everybody. If the number grows to a dozen or more it becomes impossible.

There are danger signals when the quality of membership sinks. When two or three of the bosses are regularly sending their subordinates instead of attending themselves, the committee is clearly on the wane.

The third danger signal is in the aura of the meeting itself. An experienced observer entering any meeting quickly gains a sense of the pace and productivity of that meeting.

Suppose that danger signals are perceived. Suppose that

172

the chairman evaluates and recognises that a committee is no longer as productive as it has been. What can he do?

He has three options:

Cure

Resurrection

Death

CURE

Ways to try to cure include:

Change the frequency of the meetings

Change the terms of reference

Limit the time

Change the membership

Change the chairman

Cut the members

These are all palliatives. Each of them can impede the decay of a long-standing committee, but they will not cure it.

They may enable it successfully to complete a task stretching over another half-dozen meetings, but they cannot totally avert the inevitable processes of decay.

RESURRECTION

The process of resurrection means accepting that the present body has atrophied. Suspend it and create a new body. Give it the same task but fresh membership, fresh chairman. Tell them to work out their own ways of operating; and look for tighter timing or more positive results.

DEATH

Difficult. Inevitably the committee has become the hobby horse of some people. Older members of the organisation, possibly some who have been influential, but are no longer so productive, have found this to be a useful retiring ground in which they have felt they can make some contribution.

There may yet be some need for the task and objectives which were the committee's terms of reference, even though it does not justify the full amount of time and effort which it is commanding. Nevertheless there is a reasonable excuse for keeping it going.

It is not businesslike. Once it is into the decaying phase, a committee should be helped quietly through its death throes.

Delay decision on 'Date of next meeting', and let it die quietly; or be more firm and give some individual the task originally handled by the committee.

Summary

Repetitive meetings go through a sequence of predictable phases.

Creative

Regularising

Productive

Decaying

There is an inevitability about moving through the exciting developmental, towards the routine and ultimate decay into the ritual.

In the early phases, the sequence of meetings absorbs energy and competence.

Subsequently yesterday's problems and yesterday's means start to sap today's energies.

Decay may be arrested but not prevented.

Resurrection/euthanasia are the options.

21

Bigger meetings

The techniques needed by chairmen and members of bigger meetings are in principle along the same lines as the techniques already discussed for business meetings, but there are a number of differences, and this chapter will be devoted to them.

The techniques are here considered in two separate groups:

Conference chairman

Conference member

Conference chairman

Before the meeting the conference chairman should have done his homework. He should know the topics to be presented, he should know who is to present them, he should know the time available.

He should also have considered the possible form of discussion and should have prepared sufficient prompts to be able himself to vitalise discussion should it be flagging after the presentation of a paper.

During a large meeting, the chairman's role is in four phases:

Introduction of speaker

Behaviour during presentation

Discussion-leader

Ending the session

INTRODUCTION OF SPEAKER

The chairman should ask the speaker, either well in advance or immediately before the meeting, how he wishes to be introduced. Ask him to write his name and qualifications so that no mistake is made. Preferably ask him for a typed statement of how he wishes to be introduced.

In introducing the speaker, the typical errors are the obvious ones. The first is that the chairman gets the speaker's name wrong. Instead of introducing 'John Dennis' he peers through his spectacles and says 'Our speaker to-day is John Davies'. To the consternation of Mr Dennis and of the members.

The second error is to make the introduction inordinately long. Some chairmen even go on for five minutes, extolling the speaker and at the same time throwing in a few of their own choice hobby horses. This helps neither speaker nor audience. Anything beyond a couple of pithy paragraphs is unreasonable.

The third regular error is to steal the speaker's thunder. The chairman is not content simply with introducing the speaker; he has to introduce the subject. He may even make guesses about what the speaker is going to say, not only stealing his thunder but possibly foisting on to him unwanted content. He may even force the speaker to disclaim or contradict a remark made by the chairman.

More appropriate behaviour is briefly to:

1. Greet the audience
2. Introduce the speaker
3. Give his essential qualifications for the task, plus any further comment which the speaker has requested.
4. State the title
5. Shut up

For example:

'Good afternoon, ladies and gentlemen. The opening paper this afternoon is to be given by Mr Arnold Jones, who is a Fellow of this Institution, a senior partner in Arbuthnot and Jones, with 27 years' experience in all aspects of the treatment of sewage. His topic this afternoon is "Disposal of sludge". Ladies and gentlemen, Mr Arnold Jones.'

DURING THE PRESENTATION

During the presentation, the audience is influenced by the chairman's attitude. His attention should be seen to be towards the speaker, riveted on the speaker. He should turn his chair towards him. His eyes should focus on the speaker. With occasional nods, smiles and eyebrow raising, he should display his concentration and his interest. It is a powerful influence for the good, helping the speaker to command attention, helping the listeners to focus energy. Yet I have seen professional engineers chairing important meetings, heads slumped on arms across the table, eyes closed.

On behalf of the audience, the chairman has the role of time-keeper. If there is a tight time schedule before the presentation and the speaker seems likely to exceed it, then it is the chairman's job to help keep him to schedule. A slip of paper bearing the words 'Five minutes' pushed in front of the speaker, is an immediate impulse. So are the two other pieces of paper which he might later push—'One minute' and 'Time's up'.

What if the speaker goes on after that time is up? The chairman has a very delicate job. In the interests of his members he must prevent the speaker going on indefinitely. On the other hand, with an eminent speaker who has prepared with great diligence, it is very difficult to interrupt a presentation. It becomes impossible of course if the speaker is reading through a prepared manuscript and would not reach his peroration for another x minutes.

But within the bounds of tolerable behaviour, it is the responsibility of the chairman to look after the interests of members. It is his responsibility to ensure that a time schedule is kept.

INTRODUCING DISCUSSION

After the speaker's presentation, it is the responsibility of the chairman to stimulate discussion. This is often done by asking for questions. This is, however, a feeble way to get a discussion moving.

Here are some alternatives which have been found to be useful:

1. Thank the speaker and ask for a couple of minutes' quiet while he takes a breather and while members set down any comments or questions they wish to put. This 'gathering period' has a positive effect on the fertile flow of questions subsequently.

2. Enquire how many people have questions or points they want to put. Instead of the normal pattern of gaps before questions get under way, one or two hands now go up; two or three others hesitantly follow; then suddenly people start worrying about being left out and everybody wants to get in on the act. The question period then becomes extremely lively.

3. The pattern of successive members getting up and asking questions, with the speaker bobbing up and down to answer each question, provokes a pattern of one-off questions and answers. It does not provide a pattern of successive development of a topic. A more positive development of discussion takes place if the chairman uses the fish-bowl technique.

   Identify half-a-dozen members who want to participate in the subsequent question/discussion session. Get them to gather in a semi-circle with chairman and speaker at the front of the hall. Then lead a question and discussion session amongst this group.

   Not only does this technique provide for a more positive discussion within the semi-circle, but also it provokes the silent majority. They suddenly feel jealous of their colleagues who have got this prerogative of talking and they feel forced to intervene. The discussion widens and intensifies.

4. When there are a number of people concerned to participate in a discussion, the chairman does well to group questions, so that all questions or contributions on one aspect of the subject are taken at the same time. Consider, for example, such a question as:

178

'Can the speaker tell us anything about such treatment of effluent when there is a high oxygen content?'

One chairman might pass the question immediately to the speaker. Another might try the grouping technique:

'Thank you, sir, an interesting question. Before I put it to the speaker, may I ask if there are any other questions on the chemistry of the matter.'

And so, the chairman should play a heavy role in stimulating and controlling the discussion period.

At the end of the session, he should do three things:

1. Thank the speaker
2. Comment on his success
3. Shut up

Conference member

Anybody going to a conference has the opportunity to be visible: to gain visibility for his organisation, and to gain visibility for himself.

Few engineers make very heavy use of this opportunity. There are always a few confirmed conference-goers who intervene in every discussion, however little real contribution they have to make, but far more engineers seem to take the opposite line. They are hesitant to raise questions or to express themselves in a large gathering.

A little practice and experience here are wonderful for creating confidence. Here are a few guidelines to help to ensure good visibility while building that confidence.

Open by stating your name and organisation.

Do so clearly.

Do so with a rising inflexion.

Avoid the tailing off—

'MY NAME is Jones of dartmoor engineering.'

Use the rising inflexion—

'My name is Jones Of **DARTMOOR EN-GINEERING**.'

Phrase the question sympathetically.

> Critical or clever questions irritate both speaker and other members of the audience.
>
> They are seen to be immature.

State the question clearly.

If a lengthy question or contribution, structure it.

> Even as little as two minutes is, for these circumstances, lengthy.
>
> Structure with title and signposts; develop question or discussion, and summarise.

Gain audience-contact.

> Use eyes, voice, posture and gesture to influence the whole room, while nominally addressing the chairman or speaker.

Be careful where you sit.

> It is impossible to have reasonable audience-contact from the front, the middle or the back of a theatre-style auditorium.
>
> The best place is at the end, either left-hand end or right-hand end, about three rows from the front. From this position the questioner can command contact with the remainder of the room.

Summary

In bigger meetings both chairman and speaker need to make use in principle of techniques described for smaller meetings, but they need to be employed in slightly different ways.

The chairman should:

Do his homework.

Find out how the speaker wants to be introduced.

Introduce the speaker briefly.

Focus his attention and concentration intensely on the speaker.

Be creatively stimulating in subsequent discussion.

The ordinary member should:

Look for reasonable visibility.

Have a high opening impact.

Be positive in his questioning.

Make contact with all the audience.

Sit in the best position to do so.

Part 4

Effective interviews

Introduction to Part 4

For this part of the book, an interview is defined as a face-to-face meeting, formally, without control by a chairman. This definition excludes informal dialogue.

Most interviews are one-to-one: that is, only two people are present. Occasionally one side has more members, but that special situation is not part of this consideration.

This part opens with a general chapter on interviewing.

In some interviews, one party is dominant, for example in selection and appraisal interviews. There is a different situation if both parties have equal power—the situation assumed for a negotiating interview.

Distinctions between those types of interview will be made in separate chapters, respectively on selection, appraisal and negotiating.

To complete the book, there will be a final chapter 'On being interviewed'.

22

Interviewing

This general chapter on interviewing considers:

1. Complexity of the situation
2. Preparation of thinking
3. Preparation of the setting
4. Opening the interview
5. Style of interview

Complexity of the situation

Many people approach interviews in a state of tension and uncertainty. They feel that the situation is a threatening one, possibly one to which they are not accustomed, and possibly one in which matters of great significance will be weighed in the balance.

These concerns are reinforced by uncertainty about what to expect. There is no universal precedent about the form which an interview will take. There may also be uncertainty in a person's mind about the amount of time it will last, and about the criteria on which momentous matters may be judged.

Such concern and uncertainty is not likely to be felt by the dominant party, if there is one. He will probably be more accustomed to interviews; he will dominate the procedure of discussion; he will be the chief determiner of time span.

The concerned party is anxious to project a particular image, trying to ensure that unappetising information is not sent; and he filters also the message which he receives, interpreting it in line with his own expectations and beliefs. The fraught conditions are all too likely to lead to poor communication.

In practical terms, the barriers facing the interviewer are:

1. He is only half heard. The interviewee is so concentrated on framing responses or questions for subsequent discussion that he does not totally attend to what is being said to him.
2. That which is heard is ill understood. The filters go into action and the interviewee forms some picture different from that which was intended.
3. That which is understood is not necessarily believed. People are sensitive at interviews, and many are prepared for some degree of bluff and counter-bluff during an interview. Anybody so prepared inevitably introduces an element of doubt into his own acceptance of a message. He is frankly suspicious.
4. The messages sent to him are likely to be decimated by nervousness, or may be selected to impress or to justify, possibly omitting the supposedly unpalatable.

To overcome these barriers, the interviewer needs to use techniques skilfully which he can use only if he is well prepared.

Preparation of thinking

The need is to establish sharp, clear pictures of the information to be exchanged, and of the means to do so.

The method of preparation here recommended is the repeated use of the three-stage technique first outlined in Chapter 2 and described as the A4/A5/A6 technique. To repeat in summary form:

1. The A4 stage is the setting down of random thoughts. The purpose is to clear the mind of the clutter of ideas about the topic. The method is random jottings in a concentrated brainstorming period of two or three minutes.

2. The second stage is the one of getting ideas into order. The thought process is to start by identifying the key issues, normally under some four headings, each with its small group of sub-headings.

3. When the topic has thus been analysed, the final stage is the A6: the postcard, with one word printed for each headline.

Such an approach can be used repeatedly, to produce the outlines of each of the three topics an interviewer should prepare:

First, the headlines of 'information to get'

Second, the 'information to give'

Third, the 'process' plan for the sequence of the interview

There will be times when an interviewer finds one or other of those topics to be so obvious that he does not need to prepare so carefully. Nevertheless, it is a useful habit. It is all too easy to skimp competent preparation, and one often finds that the discipline of writing it down confounds one's expectations. The expected clear picture is found to be a murky mess.

Specific suggestions for the issues to be included in different sorts of interview will be given in subsequent chapters dealing with selection, appraisal, and negotiation.

Preparation of the setting

What sort of arrangement does the interviewer want for his office?

There are plenty of pitfalls. For example, there is the Headmaster who sits behind a large desk, back to the window and to the sunshine, seated in a large chair, his visitor in a smaller chair on the far side of the massive desk. (It sounds apocryphal, but I have experienced precisely this at my son's school.)

The result, of course, is to set the visitor at major disadvantage. The light shining into his eyes is reminiscent of third-degree methods of interrogation. The massive desk is both a psychological barrier and a symbol of the distance between the parties. The difference in chairs emphasises the Head's power. It is a setting which cows parents and children alike, not to mention other teachers. It symbolises an incompetent communicator.

Better practice is to:

Reduce the barrier of the desk.

Ensure that the visitor is not thrust into unfavourable light.

Sit at an angle to the visitor, not head-on.

To reduce the powers of the desk, some people like to get rid of all the intervening furniture. They use tables to rest papers or cups on, but keep such a table behind or alongside the parties, as in Figure 22.1. This arrangement has the advantage that both parties can turn together to look at papers or illustrations on the table—in this sense they come together. It is the opposite of the divisive difficulties when they are sitting opposite one another and both need to study a document.

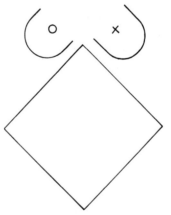

Figure 22.1 Chairs and table positioned to support discussion

This arrangement, however, is seen by more conservative people, including many engineers, as eccentric. They feel lost without some sort of table in front of them. To minimise the extent to which they are disconcerted, the interviewer has to seat them at some table or desk. It may be round or square, as in Figures 22.2 and 22.3, but it is desirable to avoid the confrontation of Figure 22.4.

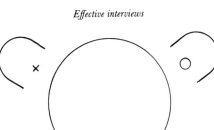

Figure 22.2 Round table, chairs positioned for co-operative discussion

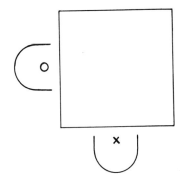

Figure 22.3 Co-operative seating at square table

Figure 22.4 All set for confrontation!

The setting must be considered not only in the sense of the furniture, but also in the 'welfare' sense. There is a need to ensure reasonable refreshment—coffee or tea or cold drinks—at appropriate times, and to forestall interruptions from telephones or from casual callers.

191

Opening the interview

At the outset the interviewer should create an atmosphere in which a valuable dialogue can take place. The characteristics of such a climate are:

Relaxed. Take the tension out of the situation so that the other party may co-operate and communicate openly.

Frank. Unless there is open and frank discussion between the parties they will shadow-box: they will deal superficially with issues and not reach towards a mutual understanding.

Businesslike. Moving purposefully along a planned path.

Rational. Concentrating on logical processes, minimising emotional statements and interpretations.

The possibility of such an atmosphere is established very quickly. As the visitor comes into the office, immediately there is a sense of the pace at which movements are taking place, the pace of walking and hand-shaking, the speed at which the parties talk to one another. A mood of hesitancy or of tension can be created within seconds if there is not a confident and open pattern of greetings between the two parties.

The first need is for the visitor immediately to see a posture which is upright and businesslike, a face which is open and has adequate eye-contact, preferably with a smile which is natural, not forced.

The pace of the parties as they approach needs to be purposeful. Not the slow and stuttering pace which promises a slow and stuttering meeting, nor the hasty pace which promises a rushed discussion.

The hand-shaking needs to be firm, but please, none of those vice-like grips!

Within seconds these non-verbal clues disclose the climate which the interviewer is seen to be establishing.

The interviewee, even when impressed by these opening moments, is still arriving tense and ill-adjusted for relaxed discussion. There is a need for a short period of ice-breaking discussion of neutral topics, until there is a meeting of minds between the parties. It is a normal part of the armoury of the skilled interviewer that he has some supply of such neutral topics which he regularly draws on—the weather, sports events, mutual acquaintances, even the morning's news.

It should be a routine to invest a limited time, maybe a couple of minutes, in this ice-breaking before moving into the substantive part of the interview.

As a bridge between the ice-breaking period and that substantive phase, it is helpful to give the interviewee a sense of the path planned for the interview. What is the sequence of events going to be? How long will it take?

My own method here is to leave a fair amount of initiative with the other person. I want to set him at ease. I want him to play a full role. I do not want to be seen as a domineering interviewer. And I want as soon as possible to have him talking and not for me to be doing it all.

After welcoming him, probably standing with him during the ice-breaking period, I suggest that we should sit down. As we sit down I will move towards business with some remarks on the following lines: 'Well, thank you for coming to see me this morning. We have a mutual interest in 'x' and there will be some information which you want to get from me, and there is some information which I should like to get from you. I suggest we will need about 45 minutes to do that. Is that all right for you? Would you prefer that we start by my asking the questions, or by you asking them?'

A short dialogue now leads the two parties naturally into the substantive part of the discussion, with the barriers and the uncertainties of the situation at least partially resolved.

Style of interview

Every enterprise has its characteristic style of management. There is an equally wide range in styles of interview.

Experts on styles of management have argued that there are 'ideal' styles for managing. This I find impossible to accept. The style appropriate for any organisation is a blend of its history, of its present personalities, of its situation in the market-place and in the community, and of the nature of its work. The imposition of a novel style is costly, takes years, and is often doomed to frustration.

Each style of management can be associated with a distinctive style of interviewing, and just as the style of management should vary from enterprise to enterprise, so should the style of interviewing.

Here are some examples of differences of style:

> *Formal* and *informal*—The meticulous upright interview, great attention to detail, dress and presentation. High emphasis on formal qualifications. Contrast the relaxed atmosphere and the stronger weighting on experience in the informal interview.

> *Authoritarian* and *democratic*—Dominance of the senior person in the one organisation, contrasting with the power sharing and discussion-sharing of the other organisation.

> *Bureaucratic* and *personalised*—The adherence to rules and procedures on the one hand contrasting with the individuality and the importance of the personality on the other.

Whatever general advice may be given about techniques for interviewing, I believe it is important that they should be blended to suit the style of management in each organisation.

Summary

1. At least one party at an interview is liable to be tense and uncertain, and consequently to communicate particularly poorly.

2. The interviewer correspondingly needs to have done his preparatory thinking, carefully considering the information he needs to give and to get, and the way in which this should be done.

3. The setting for the interview should encourage a relaxed and productive discussion.

4. The interviewer needs a technique for breaking the ice and for moving towards the substantive discussion.

5. The style of interview should reflect the organisation's management style.

23

Selection interviews

The topic of selection interviews is treated in five sections:

Preparation for interview
Conduct of selection interview
Discussion-leading
Choice of interviewer
Assessment of interview

Preparation for interview

In moving towards choice of a candidate for appointment to an important position there are four documents which need to be prepared:

1. The job specification. An internal document defining job title, organisational relationships, duties and responsibilities, terms and conditions of appointment.

2. Person specification. A statement of the job requirements, in terms of the person needed to fill the appointment. I find it helpful to build this specification under four headings.

    Calibre: What is the essential level of qualifications/experience/mental or physical ability? What academic or other level of qualification is needed? (Do not over-estimate!)

> Technical ability: In what subjects or pro-
> fessions or trades? What qualifications?
> What experience?

> Personality: With whom will he have to be in
> contact? In what ways is it important that
> he should fit in with others? Will he have
> to meet others or is there need here for an
> individual whose best work is done in rela-
> tive isolation?

> Motivation: Self-starter or externally trig-
> gered? Good at routine or good at dra-
> matic? Concerned for power or
> achievement or affiliation? Balance of
> interests between work and leisure? Need
> for approval.

These four sections of the person specification can later be incorporated into a fourth document, the interview assessment. An example of a 'person specification' is Figure 23.1.

3. Job description. This document is for the information of candidates. It should contain a brief description of the company, together with the job title, organisational relationships, and a summary of the duties and responsibilities of the job.

4. Interview assessment form. The contents of this form will be dealt with in a later section, 'assessment of interview'.

In addition, the interviewer must prepare his own thinking in some such manner as that suggested in Chapter 22.

Conduct of selection interview

Some things will make themselves apparent during an interview, however it may be conducted. For example, regardless of the topic being discussed, the quality of a candidate's thinking becomes apparent from the answers which he gives to questions, and from the way in which he himself frames questions.

It may be desirable to check those impressions. In part that

PERSON SPECIFICATION
_____

ENGINEER – MINING
_____

CALIBRE   Good Honours Degree

                   Fully qualified professionally

                   Able to describe and discuss project in strategic
                     terms

TECHNICAL   Qualified Mech. Eng. and Mining Eng.
 NEEDS
                   At least 5 years' experience in major mining projects

                   Balance of field and design experience

                   Experience in client liaison

PERSONAL   Must acquire confidence of:
 RELATIONSHIPS
                     Client's representative

                     Own design staff

                     Colleagues in other departments (NB Survey Office)

                     Sub-Contractors

MOTIVATION   Will be required to start and sustain impetus on
                     projects

                   Will need high tolerance for eccentric contacts
                     and concern/capability to convince them.

Figure 23.1 Example of a 'person specification'

can be checked from the application form, by finding the extent to which a candidate has succeeded in building qualifications. Other tests are also possible. Some people rely heavily on intelligence tests, but there are others (I am one) who have found such tests to give a misleading impression of the way in which a candidate will apply his mind. Intelligence tests are imperfect.

197

Other tests may help in the assessment of manual or physical dexterity, if these are job requirements.

Whether intelligence tests are used or not, there is little need to slant the direction of an interview simply to judge the candidate's calibre. Equally, the interviewer will find that he receives messages about a candidate's personality without framing questions in that direction. From what he sees and from what he hears, regardless of the topic being discussed during the interview, the candidate's calibre and personality become visible.

Effective interviewing therefore focuses on the other two issues: technical competence and motivation.

Technical competence is shown in part by qualifications and experience, but there is always need to probe. In his exams, what were his best subjects? And his worst? What are the details of his experience? In what ways has the candidate had relevant experience previously? In what ways has he not had corresponding experience recently? What evidence is there that he can (or cannot) overcome the gaps in his experience?

Motivation is more difficult. There is a school of thought which believes that a person's behaviour derives from heredity and the environment in which he has grown up, and that systematic analysis of this background leads to an accurate breakdown of how he will behave. This leads some interviewers to go in for a 'biographical' interview, finding out all they can about the person's family and upbringing.

This is not a technique for any but the most specialised of selection interviewers. It requires great knowledge of the type of information needed, and the establishment of a great deal of trust with the candidate in order that he will reveal it. Even then it is very difficult to interpret.

For the less specialised interviewer there is more to be said for building an understanding of what has and what has not motivated the candidate in his previous work. What has he most enjoyed doing? What has he least enjoyed doing? What were the characteristics of the work and of the environment at that time? What sort of boss did he have? Who was he working with? What were his degrees of dependence/independence? What does he say are the characteristics he is looking for in the job?

Discussion-leading

In selection interviews, the interviewer needs to get information.

He gets information when he is listening. He is not getting information when he is talking. It is a mark of the unskilled interviewer that he tends to talk too much, possibly taking 60% or even 70% of the interview time in talking himself.

The really skilled interviewer listens for the majority of the time. He can cut his own statements and questions down to 40% or even 30% of the interview time.

A technique helping in this direction is the use of ballpark questions: that is, questions offering the interviewee a wide range of options about the way in which he will talk. At a selection interview, for example, a ballpark question might be: 'Well now, will you tell me about your experience for the sort of appointment we are discussing?'

The interviewer can sit back and listen, using the techniques of effective listening described in Chapter 7 of this book. He will find out a great deal about the interviewee's relevant experience. He will at the same time find out about the interviewee's judgement of what is relevant.

Not everything. The candidate from his own frame of reference cannot be expected to talk about everything which may seem necessary from the interviewer's frame of reference. It is up to the interviewer to follow up the ballpark questions with one or two detailed questions in due course—questions which should some readily to mind as a result of the preparation which he has done in advance of the meeting.

In framing these questions, he should use open-ended phrasing. Compare the following examples:

> 'Will you have difficulty in getting to work here?'

with

> 'What will be your journey if you are coming to work here?'

> 'Have you experience of this sort of work?'

with

> 'What experience have you of this sort of work?'

Positive listening is a mark of the skilled interviewer. The interviewee sees and indeed feels the concentration of a skilled interviewer. It is visible in the way he looks, in the patterns of his eye-contact, in the way he sits. He radiates energy while he listens. He needs to show that he recognises points as they are being made. He smiles, signals curiosity with his eye-brows, surprise with his gestures.

He uses these visible signals to encourage dialogue just as much as he uses the spoken word—indeed the skilled inter-viewer makes even more use of the non-verbal than the verbal elements of communication.

He is also alert to non-verbal signals from the other party. For example, the point at which the interviewee is making borderline statements—ones which might be the truth but may not be the whole truth—are normally signalled by aver-sion of the eyes. Provided always that the interviewer is alert for such signs, he will notice many which are perfectly obvious.

The interviewer's job is not only to get information, but also to give information. The way in which this is done can influence the candidate's interest in accepting or refusing the job offered. Even more directly, it can influence his later atti-tude and commitment to the job.

The job description should of course give the candidate a formal understanding of what the job is and what it will need. But he should have more than that: he should have the infor-mal sense of what it will really be like to carry out the job. The interviewer should give that sense, warts and all. Each job has desirable features and other features which are less desirable. The interviewer should not shrink from making the candidate aware of the less desirable—otherwise when he comes to take up the appointment his disillusion will be strongly demotivating.

The interviewer should of course be sensitive about defining the less desirable aspects. If the reality is stated to be 'You will never get the drawings back from the next office on time', the candidate gets a quite different impression compared with 'There are the normal challenges of getting drawings back on time from the neighbouring offices'.

Previously, we recommended the interviewer to let the can-didate lead in the interviewer's giving of information, to let the candidate put the questions. He cannot of course be expected to put every possible question, and it remains the duty of the interviewer to ensure that his visitor goes away fully informed.

Choice of interviewer

Who should conduct the selection interview?

Normally several people should participate in interviewing, but not all together. An interview is a face-to-face dialogue and the thoughts of different interviewers inevitably move along different tracks. It is not possible to get a dialogue in depth with a candidate if discussion jumps from track to track—from interviewer to interviewer. The need is for that degree of continuity which comes best when there is only one interviewer at a time.

The people who normally should conduct interviews include:

1. The *boss*. He who will be responsible for the candidate, if appointed. A great deal of any individual's effectiveness depends on his relation with his boss. None of us is perfect; each of us performs best when supported by a boss who feels some commitment to us. That sort of commitment comes from having the responsibility of having hired the candidate. A mediocre candidate can perform well if the boss feels commitment to him, where a better candidate might fail with a less committed boss.

2. The *technical expert*. One who has a speciality for the job—a specialist engineer for such an engineering appointment; an accountant for an accountancy job, etc.

3. The *personnel specialist*—experienced in probing motivation and the personal elements of appointments. One also who is experienced and skilled in discussing terms and conditions of the appointment.

Assessment of interview

The assessment of an interview should be done with the help of an interview appraisal form. This should contain informa-

201

tion on each aspect of the 'person specification': the calibre, technical competence, personality and motivation of the candidate.

I find it helpful to write down the facts opposite each of these four headings as far as I have been able to glean them through the interview. Then to put some rating against each. I use a 0–25 scale on the criteria: 5, unsatisfactory; 10, barely satisfactory; 15, satisfactory; 20, good; 25, outstanding.

Having rationally evaluated the candidate I leave scope for the element of 'hunch'. This is part of what every interviewer inevitably feels sometimes. Despite the excellence or otherwise of a candidate on all the rational grounds which one has sought, one sometimes gets some other feeling for better or for worse.

I am prepared to raise or lower the rational total by up to $+5$ or -5 for this 'hunch'.

An example of a completed interview assessment form is Figure 23.2.

The interview assessment form is of particular help when the three interviewers meet at the end of the day. It is remarkable how often they are in agreement. When they are not, differences must of course be discussed, and the form then helps to pinpoint the differences.

If differences cannot be fully resolved, if there is some difference of opinion rather than of fact, then it should be the views of the boss which predominate since it is he who will make or mar the success of the chosen candidate.

The selection interview, as here discussed, is of course only a part of a full recruitment procedure, which should also include a means by which candidates are attracted, the possibility of using tests, and the extent and means by which references are followed up.

Summary

Selection interview: preparation should cover—

    Job specification

    Man specification

    Job description

    Interview assessment form

INTERVIEW ASSESSMENT FORM

JOB - ENGINEER - MINING CONTRACTS

CANDIDATE *I. N. DIVIDUAL*

	SPECIFICATION	EVALUATION
CALIBRE	Good Honours Degree Fully qualified professionally Able to describe and discuss project in strategic terms	*1st / Sheffield* *FIMechE* *Success in design office* *Highly competent* 5 10 15 20 (25)
TECHNICAL NEEDS	Qualified Mech. Eng. and Mining Eng. At least 5 years' experience in major mining projects Balance of field and design experience Experience in client liaison	*Fully qualified* *6 years' experience in mining* *Bulk of experience office-based and experience is that* *'Clients were fools'* 5 (10) 15 20 25

PERSONAL RELATION-SHIPS	Must acquire confidence of:
	  Client's representative
	  Own design staff
	  Colleagues in other departments
	    (NB Survey Office)
	  Sub-contractors
		Assertive and abrasive manner
		Strong talker; no listening skill
		Evidence of poor previous working relationships ('their fault')
		(5) 10 15 20 25

MOTIVATION	Will be required to start and sustain impetus on projects Will need high tolerance for eccentric contacts and concern/capability to convince them	*Ambitious person, working hard over long hours* *Imaginative and creative, but low in tolerance for others* 5 10 (15) 20 25
COMMENTS	*Could not handle the relationship demanded by this job.* *Superb candidate for a more individualist / isolated role.* *Have we any coming up?*	TOTAL ABOVE: 55 HUNCH -5 (0) +5 0 TOTAL 55

Figure 23.2 Example of completed interview assessment

Each interviewer should make his personal preparation for the information he wishes to get and to give.

Selection should normally involve three interviewers: boss, technical specialist, personnel specialist. They should interview independently of one another.

In arriving at a final decision, the views of the prospective boss should dominate.

24

Appraisal interviews

The case for conducting appraisal interviews can be strongly made. Each individual has the right to know where he stands in an organisation, how his endeavours are viewed, what are his prospects for the future. Every boss has the responsibility of counselling his subordinates and helping them to improve their performance.

Simple and obvious as it sounds, it is by no means so simple in practice. When boss and subordinate are working in day-to-day contact, they develop a mutual esteem and, it is hoped, an intimate partnership. It is difficult for both parties to stand back from that close partnership and to discuss personal competence. It is the sort of discussion which can often change a relationship—by no means always for the better.

Appraisal interviews may be favoured on the grounds that they should help a person to modify his behaviour. In practice, it is not easy for people to do so. If a person is a naturally abrasive character, he is a naturally abrasive character. It does little good to tell him so—and it may do a lot of harm. It may undermine a person's confidence, make him over-sensitive, and hamper his relationship with the appraiser.

There are those who believe that such people should be appraised and steps taken to help them to change their behaviour. However, few of us have the competence so to change our own or other people's behaviour, and ethically it is questionable whether it is desirable to change another's character. At best it is risky.

The conduct of an appraisal interview is therefore a sensitive matter, deserving special treatment and consideration. In this chapter the topic is considered under the headings:

Style of appraisal

Preparation for appraisal

Conduct of appraisal interview

Follow-up

Style of appraisal

There is no uniquely appropriate style for an appraisal.

Different organisations have radically different patterns of behaviour. The individuals within each organisation adapt to that pattern. They adjust their own behaviour accordingly, and they accept that the recognised way of doing things will be sustained.

It is particularly important that such a sensitive occasion as an appraisal interview should fit into that known and accepted pattern.

If, for example, the style is militaristic, with a clearly defined hierarchy of seniority, in which bosses tell subordinates precisely what is required of them, the appraisal interview should be of the same form. Boss should tell subordinate precisely where he stands.

In a bureaucratic organisation, where written rules and procedures govern conduct, the appraisal should bear the marks of bureaucracy. Heavily stylised, certainly written, checked and counter-checked by those in authority.

In a highly personalised organisation—for example, a small consultancy headed by a powerful character—the individuality of the boss permeates the whole establishment. His word and his way of doing things carry a charisma felt by all. He and his style of working should characterise the appraisal interview and he should not let his methods be diluted by any expert advice to the contrary.

The form of an appraisal interview therefore needs to reflect the style of the particular organisation. The whole system of appraisal should be worked out to suit that style.

Preparation for appraisal

Question one is 'Who conducts the appraisal interview?'.

There are three main candidates:

1. The person's boss
2. The boss's boss
3. An independent 'expert' such as a personnel officer

The person's boss should have the most intimate knowledge of the individual. To that extent, he should be the person best equipped to conduct an appraisal interview. There are, however, disadvantages: the sensitivity is such that working relationships could be prejudiced. Moreover, the proximity between boss and subordinate can inhibit him from standing at sufficient distance to take a dispassionate view of the individual's performance.

The boss's boss (the 'Grandfather' in the organisational hierarchy) is favoured by some experts. The grounds are that his distance enables him to have a better all-round perspective; and that his seniority signifies concern that a subordinate should be well recognised. The seniority is said to enhance the credibility of the appraisal interview.

There are stronger disadvantages. If the organisation is a formal one in which the 'Grandfather' really does carry such prestige, then he will have little direct contact with the subordinate. On the one hand, this will turn the interview into a frightening experience. On the other, the interviewer will inevitably be influenced in what he says by the intermediary, the boss, and may be forced to give and to defend views not entirely his own.

Purely on practical grounds, the use of 'Grandfather' in this role is also suspect. If each boss has half-a-dozen subordinates then each grandfather has 36 grandchildren. Proper preparation for and conduct of 36 appraisal interviews would command an excessive amount of his time.

The third option is to have appraisal interviews conducted by 'experts'. This is not a commendable practice. It inserts outsiders into the key relationship between boss and subordinate. It inserts people who cannot be aware of all the relationship's nuances, or of the problems and achievements of the individual within his working situation.

My recommendation therefore is that appraisal interviews, if they are to be held, should be conducted by the person's boss. It is he who is responsible for the output of the subordinate; he more than any other who can create conditions in which the subordinate can best perform; he who may have the most to learn in any appraisal interview.

How should he prepare himself?

There are many different approaches to the subject of appraisal, but they can all be classified within one of four types: traits analysis, performance analysis, results appraisal and strengths analysis.

Traits analysis Consider such traits as technical competence, commitment to work, relationships with superiors, leadership, etc.

> These are subjective issues. The boss's judgement of them is called strongly into question. Where this system is used, there is need for some check on that judgement—for example by the boss's boss.

Performance analysis Considering each responsibility of the individual described by his job specification, and rating performance which is good, bad or indifferent.

> This procedure can be used mechanically, reverting at the time of each appraisal to an established job specification. Used in this way, the process becomes mechanical and drifts away from a meaningful assessment.

> More practically, it may be recognised that people's jobs are constantly changing. The environment in which they work changes; their colleagues, boss and subordinates change; the individual increasingly imposes his own character. The job thus is constantly changing.

> Boss and subordinate are helped together to be more effective by regular discussions of what the job really consists of *now*, and by joint examination of the difficulties encountered in doing it.

> This commendable and creative use of performance appraisal has two drawbacks. One is that it is heavily time-consuming, both for boss and for subordinate. The second is that it demands a degree of articulacy and self-analysis which does not come easily to everybody.

Results appraisal This approach was heavily developed within the concept of 'Management by Objectives'. It is characterised by standards and targets for achievement in the next phase being discussed and agreed between boss and subordinate. Criteria for evaluating results are agreed at the same time and subsequent appraisal is on results achieved in those terms by the end of the period. Discuss and re-cycle.

> Successful use of this approach demands a style of managing in which much is delegated, and in which the individual is encouraged towards setting his own targets.

> Excellent in theory, in practice it all too readily degenerates into a paper exercise, with the paraphernalia and the paper store being treasured more than the dialogue between boss and subordinate.

Strengths analysis This is an original approach to systems of appraising. It is very simple. Basically it concentrates on two main questions.

> 1. What are the three or four things which the individual is very good at?

> 2. What is one thing which causes him the greatest difficulty?

> This analysis, concentrating on strengths, is a positive starting point for an appraisal interview. It is also positive for both boss and subordinate in stimulating the development of the individual's role and/or of his career.

The boss working towards an appraisal interview will normally find that he is 'helped' by one or other of these systems. He then needs to sharpen his thinking and to decide on the key issues for discussion. This stage of preparation has been discussed previously, in Chapter 22.

Conduct of appraisal interview

The basic ground rules apply. Do your homework. Start with a firm and simple understanding of the key points which need to be made. Take time on ice-breaking. Bridge the development towards the formal agenda with comments on how the meeting is to be conducted.

Conduct the discussion in ways which are natural for the style of the organisation. Let it flow naturally as in other discussion between boss and subordinate.

Subject to the constraints imposed by that natural behaviour, here is a commendable sequence for the main body of the appraisal interview. It is in four stages:

Exploring

Informing

Discussing

Deciding

In the *exploratory* phase, the interviewer is concerned first to establish a pattern of creative thinking with the subordinate. This is done, not by imposing the boss's ideas and impressions, but by seeking first to find out the subordinate's.

Most of us are self-critical. We rate ourselves lower than many others would do on many traits. Provided we feel confidence in our bosses we are happy to explore with them and get from them helpful comment and constructive criticism, but we want to start with a feeling that our self-interest is significant.

In the exploratory phase of an appraisal interview, therefore, the boss seeks to find out the subordinate's views on:

How the job is going

What have been the successes

What have been the disappointments

What support is required

The next stage is the one of the boss *informing* the subordinate of how his performance is perceived. Hopefully, 80% of this giving of information will be confirmation of the subordinate's own views. The remaining 20% should be offered, clearly and firmly and objectively, as the dispassionate view of the outsider; both stressing the strengths and achievements, and recognising the difficulties.

In the *discussion* phase the key concern should be with performance improvement. To some extent this may mean a change by the individual in the way in which he works. Often to a large extent, performance can be improved by the

210

opportunity to perform better. This may be in the form of changing responsibilities, or changing relationships, or the boss opening avenues previously seen to be closed. Creative discussion between the two should pave the way for fresh action.

In the final phase of *deciding*, the boss and subordinate together should decide what action needs to follow the appraisal. The most likely actions will fall into four categories:

1. *Changing priorities.* Modifying the priorities which the subordinate is giving to the different aspects of his task.

2. *Training.* Steps to help personal development, whether by special projects, by experience in a new form, or by training courses.

3. *Conduct.* Changes of behaviour suggested by the discussion, and especially steps needed to improve any difficult relationships. The boss himself has a key role to play when there is a need to bridge a difficult relationship.

4. *Job conditions.* Modifying the responsibilities or the relationships in the job specification.

Follow-up

The product of a successful appraisal interview should be twofold.

First, it should enhance the mutual understanding between boss and subordinate, leading to improved mutual support.

Second, there should be some specific action points. Decisions should have been taken about priorities, training, conduct and job conditions.

It is the duty of the interviewer to follow up this intended action and to ensure that it does take place.

Summary

1. The style of an appraisal interview should conform to the organisation's style of working.

2. The best person to conduct an appraisal is the individual's boss.

3. There is a variety of procedures designed to help appraisers:

> Traits analysis
>
> Performance analysis
>
> Results appraisal
>
> Strengths analysis

4. Appraisal is a sensitive form of interview. Basic ground rules apply:

> Do homework
>
> Break the ice
>
> Move gradually into the substantive part of the discussion.

5. If organisational style allows, the sequence of discussion should be:

> Exploring
>
> Informing
>
> Discussing
>
> Deciding

25

Negotiating interviews

This topic is dealt with under three headings:

Styles of negotiation

Structure of a negotiating interview

Preparation

Styles of negotiation

There is a fundamental distinction to be made between two styles of negotiating: the creative style and the competitive style.

Favourable conditions for a creative negotiation include one in which both parties are equally interested in reaching an agreement, in which they are not concerned to exert power, and in which the culture of the two parties favours a creative style.

Given such circumstances, the style of negotiating becomes open, frank, creative. There is readiness to give information and to share initiatives in the joint interest of negotiating towards agreement.

In other circumstances negotiations become competitive, tough, with each party concerned to gain advantage for itself. A series of outposts become established during the negotiation from which the parties later withdraw to more heavily pre-pared stockades. This competitive form of negotiation is dif-ferent in character, in practice and in attitude.

Distinctive skills are needed to handle each type, and it is not productive to impose any one style on all negotiations.

Some people are good at doing things one way, some at another. The advantage lies with the one who does what comes naturally, and who also takes the trouble to check and polish his negotiating skills.

It is impossible for him to do so in the negotiating room. A book (such as this) may help him get a few impulses, but is no substitute for a practical training seminar.

Structure of a negotiating interview

Whichever style of negotiating is adopted, the foundations of establishing a proper climate and of agreeing on procedure should be constant elements.

The skilled creative negotiator starts by trying to establish at the outset a climate which is brisk and businesslike, cordial and co-operative. He invests a lot of time in ice-breaking.

The competitive negotiator, on the other hand, aims to be brisk and businesslike, but at the same time is tempted, at the outset, to probe for his opponent's strengths and weaknesses. From an early stage a defensive/aggressive mood is being created.

Shortly, the negotiations need to move from informal chatter towards substantive discussion. This movement can productively be bridged by agreeing on the purpose, plan and time available for the negotiation. The initiative in making these procedural suggestions usually lies with the host, but the 'pecking order' is not of much concern to creative negotiators. Their goal is mutually advantageous agreement, not self-assertion.

The competitive negotiator, on the other hand, seeks to gain advantage in the way in which the agenda is established. He will, for example, be early anticipating areas which may become contentious between the parties; he will want to ensure that he can first discuss matters on which he expects us to concede, before getting into areas where he may have to.

These differences of behaviour within the opening minutes establish that the remainder of the interview will be conducted either creatively or competitively.

The substantive part of the negotiation follows through five stages:

Exploration

Joint assessment

Bidding

Bargaining

Settling

In many negotiations the boundaries between these stages are confused and uncertain, but the skilled negotiator has the distinction between the stages strongly in his mind. Either formally as part of the agreed agenda or informally through the way he works, he tries to keep order and to follow through the sequences in such a logical fashion.

The emphasis on the stages varies depending on the style adopted. In creative negotiations there is heavy emphasis on the phases of exploration and of joint assessment; the later stages of bidding and bargaining are then nurtured by mutual understanding and respect for respective positions and interests.

In the competitive style, on the other hand, exploration and joint assessment are different in form, and the bidding and bargaining phases are more significant.

Let us follow through the sequences for each style.

CREATIVE NEGOTIATING

Exploration starts with statements from each party about their interests in the topic being negotiated. Such opening statements should include the party's expectation about the broad areas within which the negotiation will take place; and statements of what they want, what they can contribute, what are their priorities.

It is desirable that such statements be made on a broad front without yet deep-diving on any particular issue.

The other party should respond first by asking questions for clarification but *not* for justification. The distinction is important: the challenge of demanding justification shifts the negotiation towards the competitive style, and it needs very few such impulses totally to change the character of a negotiation. Questions for clarification, on the other hand, are necessary and desirable.

Subsequently, independently, the other party should offer its corresponding opening statement. This should not be an

215

alternative statement based on amendments to the first party's position. It should be a separate and independent statement. The importance of this distinction is shown up in the next phase.

Joint assessment is a co-operative and creative stage in which the parties look to see what together they can do in their joint interests.

A phrase which is regularly used by some negotiators to lead into this phase of the negotiation is: 'Well now, what creative possibilities have we got?'.

This can be the most fertile phase of a negotiation. It depends on the prior readiness of both parties openly to state their interests, and it demands the maturity to place self-interest within the context of joint interest.

The prospect of such creativity is obscured if one party has reserved or obscured its position in the earlier exploratory phase.

If exploration and joint assessment are successfully conducted, the later stages are relatively easy.

Bidding (for purchasers the corresponding word is *offering*) is informed by previous discussion. It is also taking place in a climate within which bluff and counter-bluff would be disastrous, and the bid must be more realistic than if an outpost were being set up for a competitive negotiation. Nevertheless, the same principles about choice of opening bid apply as in the competitive situation. Since they are more important in that style, they will be handled when we discuss it below.

Bargaining is also different in form in a creative style of negotiating. It is based on mutual respect, mutual recognition of difficulties and mutual searches for a solution.

In many negotiations of course there will be details in which the interests of the two parties do not coincide. There will be need for degrees of give and take between the parties, and then the basic principle of bargaining should be adhered to: trade concession for concession. Before formally agreeing to make a concession on one issue, ensure that the other party is giving an equal counter-concession. For example, if discussion has led you to believe that you may be able to yield a little on your price, do not do so until he has offered you corresponding help with, say, delivery. The phrase to use here is 'Well, I think that it might be possible to take a further look at the price issue, but could we first of all settle this issue of delivery?'.

Tactics of the bluff/counter-bluff are not appropriate for a creative negotiation. There are, however, two tactics which are positive and which deserve regular use.

First, the use of the recess, taking a break of a few minutes during which each party moves out to reconsider the process, to re-prepare, and to freshen minds which may be getting dulled by the demands of the negotiation. Note that after any such recession, when the parties re-assemble, they need to have a mini version of the opening of the negotiation. Brief time for ice-breaking; followed by joint review of procedure before reverting to substantive matters.

The second tactic is 'the Golf Club'. When negotiations get sticky, when the climate of the negotiating room stops being progressive, when discussion reaches a point at which it is becoming unproductive, there is need for a different form of freshness. It can often be found by withdrawing to a different atmosphere, one in which trust, open and responsible behaviour are the accepted pattern. It may be found in the golf club or in the gentleman's/lady's club. (In Finland, it is in the sauna.)

And so the process of creative negotiation moves through major phases of exploration and joint assessment, and through bidding and bargaining phases towards settlement.

COMPETITIVE NEGOTIATIONS

Competitive negotiations move through the same sequence but the pattern within the sequence is radically different.

Exploration in this case is a testing and probing operation. Each party is trying to establish its own interest, to see what it can get from a prospective deal, and to see how it can influence the other party to its own advantage. Particularly, each is building information and building outposts for the later stages of bidding and bargaining.

The information may be either business or—occasionally—personal. The business level is set by questioning business climate, successes, problems.

'How is cash flow with you?'

'How are your deliveries nowadays?'

'Have you overcome your problems with quality?'

Some negotiators are taught to go on to the personal level, probing for personal problems or weaknesses, in the hope of being able to exploit them at a later stage. This is not the type of behaviour commended in this book, but it is behaviour to be anticipated from others in some circumstances. When it happens, deflect it; then be certain that your preparation has taken the form described below for competitive negotiations.

Even at this early stage of the negotiations, the rule of '*Get–give*' applies. Get information from the other party before giving any, if you possibly can. At the least, trade information for information.

In the phase of *joint assessment*, the parties begin to perceive the shape of a prospective deal between them and to identify the issues which such a deal will raise. Typically, for a substantial contract being negotiated fully (without limits imposed by earlier tender procedures), such issues would include:

The size of the deal

The speed at which it should be done

Partial or turnkey contracting

Technical requirements

Quality assurance

Financial terms and conditions

Prices and discounts

Bonds and guarantees

Tendering processes restrict the scope for negotiation, but nevertheless it is important for the parties to explore and assess the matters they will need to negotiate, and the priorities for discussing them. For example, will supplier's technical reservations be considered before or after buyer's concern about the price for a part of the contract?

The phase of *bidding* is particularly important within competitive negotiations. The key issues to be considered by the bidder are:

How much to bid?

How to make the bid?

How to respond to an offer?

How much? The bid on each issue should always be 'the highest defensible bid'. It has to be high since, once voiced, it is almost impossible to raise the bid at a later stage (or, correspondingly of course, to lower an offer). It has to be high in competitive negotiations because there will inevitably be a bargaining phase in which the bidder will have to make concessions somewhere; the other party in such a competitive negotiation would be totally frustrated if he were unable to win some concession. It must be high because it is a characteristic of good negotiators that they create expectations in the mind of the other party, that the final settlement will have to be within a range to the first party's advantage.

The opening bid must be high but it must also be defensible. It is certain that the bid will soon be tested by the other party and if they are able easily to attack and to get movement from the opening position, then immediately they have the bidder on the run. He will find he is kept on the run. He is in a losing situation.

The opening bid should therefore be 'the highest defensible bid'. How should it be put?

Clearly. So that it is perfectly plain what one is asking. Do not muddy the negotiation process with murky ambiguities.

Firmly. No hesitation, no stuttering about it.

Without reservations. There is no doubt about it—this is our bid.

Without justification. The inexperienced negotiator, offering justification for a bid, at the same time offers opportunities for a competitive other party to challenge.

Compare for example:

'Our price is £100 000.'

with:

'In our price we have got to cover our basic costs and our normal overheads and of course there are special problems of inflation and of exchange rate and the heavy research and design costs which we have been involved in, which mean that our price will have to be £100 000.'

The bid should therefore be put firmly and clearly, without reservation or justification.

How should the other party respond?

The other party will certainly want to raise some questions, but the distinction made previously still applies—the distinction between questions for clarification and questions for justification. Questions for clarification are usually constructive, as, for example, 'Is the price inclusive of VAT?' Questions for justification, on the other hand, are part of the bargaining process which should follow later, as, for example, 'Why is the price so high?'.

The process of negotiation is a complicated one and, the murkier the waters, the more difficult it is for the parties ever to near a deal. It is highly desirable to get bids absolutely clear before getting involved in the process of justifying and bargaining.

What after this clarification? There are two directions in which the other party may try to develop into the bargaining phase.

> On the other hand, they may try simply to whittle down the first party's bid.

> On the other hand, they may make their own counter offer.

The stronger tactic is the whittling down. It keeps the first party in the dark. If successful, it can force concessions—concessions which could be further attacked if, much later, a low counter offer were made.

In strong competitive negotiations therefore the second party will want to get into whittling down; the first party should resist and demand counter-offers before being drawn into bargaining, or, at the very least, a full statement of counter-position. Do not be dragged into deep-diving item by item.

And so the competitive negotiation moves forward towards the phase of *bargaining*.

Ground rules for this phase are:

1. Trade concession for concession. Ensure that any move we make receives a counter-move from the other party. For our readiness to improve quality get their readiness to accept later delivery.

2. Seek to negotiate on a broad front. Aim for 'headings of agreement' across the whole range of issues, then

220

for agreement in principle, oral agreement of detail ... written agreement ... legal contract. Distinguish from vertical negotiation—deep-diving on each successive issue.

3. It is always easier for the parties to accept a move on some point if another point is linked in at the same time. 'Well, just before we resolve this question of discount, can we consider who bears the insurance risk?'

4. Move at a measured pace. After the opening round of bidding, both parties form some reasonable guesstimate of the area within which a settlement may be agreed. There is some pace at which they can move in that direction. If either makes concessions too quickly, or too much, then the other party will exploit.

Suppose, for example, that you have bid £100, that the other party has offered £50 and that the expected settlement is £70. Do not rush in that direction.

It might seem that it could save a lot of trouble if you made the compromise suggestion: 'Look, you're saying £50 and I'm saying £100— let's be reasonable and split the difference. Let's call it £70'.

A really competitive negotiator will immediately reply 'But I can't afford £70', and from now on, the negotiation will be between his original £50 and our readiness to move to £70. He's got you on the run.

Don't risk it.

In this bargaining phase, in competitive negotiations, the parties may use a series of tactics.

Always positive are the uses of recess and 'the Golf Club', previously described. Other tactics include feints, pleading lack of authority, constant demand for justification, setting deadlines, and a whole range of increasingly dubious tactics. Engineers exposed to such types of negotiation are commended to the fuller treatment in the author's *The Skills of Negotiating* (Gower).

As the bargaining proceeds, the negotiation moves towards a settlement. Many skilled negotiators like at this stage to have two final devices; the final push and the last ha'penny.

'The final push' is one final concession which can be expected to seal the deal. Such a concession should be big enough to be a final major incentive, and its timing is critical. Too soon—the settlement will not yet be reached and yet more concessions will be needed. Too late—then either the deal is lost or the negotiation has dragged ulcerously beyond reason.

'The last ha'penny': competitive negotiators want the satisfaction of having won every scrap of the advantage. Even when another party offers an attractive final push, it is desirable to demand a final detail of concession to give them that satisfaction. Otherwise they will go away regretting that they settled too easily and they will be more difficult in any future negotiation.

The process of competitive negotiation then is one in which bidding and bargaining are extremely important phases and the negotiator must prepare himself accordingly.

Preparation

As for any interview, the general method recommended is the use of the A4/A5/A6 technique. Random thoughts analysed under a few short sharp headings; ultimately reduced to half-a-dozen words printed large on a postcard.

For creative negotiations; this technique should be used twice.

First, procedural preparation. The negotiator here should get his mind clear on the purpose of the interview, the plan or agenda which he will suggest for it and the length of time which he will propose.

Second, preparation of the opening statement. What are his organisation's needs/interests/resources/attitudes, etc. within the discussion area for the negotiation.

For creative negotiations, do not over-prepare. If the parties both make general opening statements there is plenty of scope for them to look creatively at joint possibilities. But if either goes into fine detail, the discussion becomes trapped in that detail. What is more, the negotiator who has prepared so finely finds his own thinking is trapped: he loses his flexibility.

In competitive negotiations, there is the same need for procedural preparation, but substantive preparation should take quite a different form. The competitive negotiator needs from an early stage to have clear ideas about his bids on each item. He needs to have identified what issues will come up and to have set figures to his bid on each.

For the toughest of negotiations, many negotiators try not simply to sort out the opening bid, but to anticipate a succession of three or four subsequent bargaining phases. They consider what concessions they should make in each phase; and the counter-concessions they will expect from the other party each time.

This is a highly detailed way of doing it. It is nevertheless a pattern in which some negotiators are trained, and for which the Engineer may need to develop his counter-measures.

If such measures are needed, then clearly they must be part of a major preparation process.

If the engineer should find himself in these very tough circumstances faced with new or different situations for which he has not prepared, he must break off the negotiation, take a major recess and refresh his preparation.

Summary

1. There is a great difference between creative negotiations and competitive negotiations.

2. Negotiating interviews should start with the establishment of a suitable climate: brisk and business-like, cordial and co-operative.

3. Bridging towards substantive discussion, the parties should agree on the procedure they will follow for the negotiations.

4. There are five phases through which any negotiation then proceeds; exploration, joint assessment, bidding, bargaining and settling.

5. In creative negotiations the critical phases are the first two: exploration carried out in an open and frank atmosphere to establish respective interests: leading to joint assessment of creative possibilities in joint interest.

6. In competitive negotiations, the bidding and bargaining phases are more heavily emphasised. Key words include 'highest defensible bid', firmly and without justification. Avoid bargaining before getting counter-offer.

7. In bargaining, trade concession for concession and keep it flexible.

8. Use the A4/A5/A6 technique for procedural preparations (purpose, plan, pace) and use the technique also for substantive preparation. If aiming towards creative negotiations, prepare a general opening statement without getting into too much detail. For competitive negotiations, identify issues and prepare bids and stances on each.

26

On being interviewed

'Being interviewed' is a dramatic event.

The context of 'being interviewed' implies that there is a meeting with a more authoritative person or people, and that they have the right to 'call the tune'. This is the situation at the interviews such as selection, appraisal and 'the professional interview'. It is not the situation of the negotiating interview, in which both parties can be assumed to be of similar status.

The former situation, 'being interviewed', is dramatic and is felt to be threatening. The candidate often approaches it hesitantly, unsure of how it will be conducted, afraid that he will fail to do himself justice. He knows not how.

This chapter is intended to help him. It is in three sections:

1. The uncertainties of the situation

2. Preparation for interview

3. Conduct at interview

The uncertainties of the situation

There are two sorts of uncertainty facing a candidate approaching an interview.

First, he does not know how the interview will be conducted. It may, for example, be cordial and co-operative, the interviewer doing his utmost to help the candidate relax and to do himself justice. It may, on the other hand, be tough. Some interviewers believe that they have the task of testing a candidate and his reactions in difficult conditions and they set out to provide a challenging environment.

The candidate cannot predict the atmosphere which the interviewer will create. Nor can he be sure of the form of the questions which he will be asked, or of any special tests he may be set, or of what opportunity he will have to raise questions. He probably does not even know how long the interview will last.

The first form of his uncertainty is therefore the doubts about the process to be followed. The second form of uncertainty concerns the criteria which the interviewer will use. He cannot be sure, for example, whether loyal and painstaking effort in difficult circumstances will be valued more or less than success in simple situations.

These are uncertainties which the candidate must face. There is nothing he can do to resolve them.

Recognise them.

Then forget them and concentrate on what can be done.

Preparation for interview

Concentrate in advance on getting well prepared.

The candidate cannot be sure what questions he will be asked, nor can he be sure what opportunity he will himself be given to ask questions.

He cannot be sure, but he can make reasonable predictions. He should organise his thinking so that he can give clear answers to the sorts of question which are most likely.

He will probably not be asked all that he has prepared for; nor can he expect that his preparation will cover everything which will turn up. But if he has done his homework properly, he will be well prepared for 80% of the interview.

Here, for example, are a number of questions which can be anticipated for different types of interview:

Selection interview

What experience have you for this job?

What was your most enjoyable previous job? What made it so enjoyable?

Why are you interested in this job?

What are your career intentions?

226

Appraisal interview

What have been your positive achievements in the job?

What are the satisfactory features about the employer?

Dissatisfactions about achievement and/or employment?

Development: in what way could things be improved?

Changing the role? Relationships? Self-development? What are the possibilities? What help will you need?

Professional interview

Project: if a thesis has been presented, then anticipate being asked for:

1. A brief oral summary of critical points

2. 'Lessons learned'

Experience: Highlights? Satisfactions?

Current professional affairs:

What are the key topics in professional journals over the past few months?

What do you know and think about them?

Career intentions:

What plans have you got for your career?

These are questions which are likely to come up at any of these special types of interview. There will be other questions, specific for each particular interview, which the candidate should anticipate for himself.

The process by which he does his preparation has to be one which enables him to go to the interview with his thinking clear and simple. The method recommended to achieve that simplicity is the one which has been consistently advocated through this book—the three-stage A4/A5/A6 technique, starting with random ideas, then analysing under a few (four?) headings, each backed up by up to four sub-headings; and finally reducing to postcard simplicity.

For some interviews, and especially for selection interviews, the candidate needs to have prepared both the responses he will make to predictable questions, and the questions to which he himself will want answers. Questions about the role, responsibilities, the position in the hierarchy, strengths needed, problems and opportunities to be expected, terms and conditions of employment, prospects for the future.

Such a range of questions might be analysed, for example, under the headings:

1. Technical requirements
2. Working conditions
3. Conditions of employment

The homework cannot cover everything which may come up in an interview. It can and should cover 80% and enable the candidate to approach an interview confident that he has done all that he could do in advance.

Conduct at interview

Each candidate is seen to have an aura. Interviewers recognise these auras. Generally, they cannot articulate how they recognise them, and they talk about them using vague words like—mature—credible—middle of the road—immature—muddled. When two interviewers sit together, both recognise the same aura, even though neither can specify how.

The aura is in fact built from a series of signals. The candidate is rarely conscious that he is sending them. The interviewers, rarely conscious that they are receiving and interpreting them. Yet the outcome is this significant message.

First impressions are critical in forming the aura. They are based almost entirely on non-verbal clues in the opening seconds of contact between interviewers and candidate. They are the clues of pace of walk, and of talk, stance, facial expression, gesture.

Dress is important too. There is a sense of what is respectable dress for engineers meeting in formal conditions. The candidate's concern to make a good impression should conform to those expectations; he should wear respectable clothing.

He can of course make a strong impression by dressing differently. He will make a strong impression if he wears a vivid shirt, no tie and casual outer-wear, but he will not create a strong *good* impression.

Immediately the interviewer's eye has glimpsed the candidate's dress, the pair are likely to shake hands. The candidate should be prepared for this, with his right hand free to stretch forward. Brief-case or papers should be carried in the left hand to avoid a minor hiatus in the opening seconds.

Predictably, most interviewers will seek for some sort of ice-breaking moment. Positive aura quickly surrounds the candidate who contributes positively to this ice-breaking.

Here are two responses to the same question:

(a) 'You got here all right?'
   'Yes.'

(b) 'You got here all right?'
   'Yes, the buses seem to be running well today'.

Candidate A has projected the interviewer into having to introduce another topic instantly; candidate B has offered him the chance of developing the ice-breaking a little further. Different auras are already surrounding the two candidates.

Positive first impressions lead to positive chemistry between the parties. The aura then develops through the main body of the interview. It depends on three streams of information which are being received by the interviewer.

First, on what the candidate says. His preparation should have helped him so to marshal his thoughts so that he can communicate crisply.

Second, on how he is heard to say it. If he is stuttering, nervy, hesitant, stumbling, interviewers suspect that he is out of his depth. If he is fast, forceful, pushful, he is heard to be brash, immature. If he answers confidently, at a measured pace, with an appropriate sense of deference, he is heard to add weight to his words. He generates an impression of thoughtfulness and credibility.

Third, the interviewer is influenced by the way the candidate is seen to be behaving. The signals are in the posture, the gesture, eye-contact.

To make this projection positive, the candidate has to have developed the skill of directing his energy in communications. He must identify himself with the interviewer, constantly checking visually the sorts of response he is getting, and use moderate gestures to reinforce his message.

He will be most respected if he is able sensitively to take on the role of listener, encouraging the interviewer to do much of the talking.

Summary

1. Before being interviewed, a candidate will be unaware of the procedure which the interviewer will follow and of the criteria on which judgements will be made. Recognise the uncertainty. Forget about it. Concentrate on what can positively be done.

2. Prepare. Anticipate the most predictable questions to which answers will be wanted, and formulate own questions. Get this thinking into sharp focus before the interview.

3. Self-presentation will produce an aura for better or for worse. First impressions are critical in creating an aura—impressions based on dress, gait, hand-shake and opening comments. Subsequently the aura is influenced by how the candidate is heard to speak and how he is seen to behave.

Conclusion

This book has outlined techniques and skills needed in four forms of communication:

Speaking and listening

Writing and reading

Meeting: chairmanship and membership

Interviewing and being interviewed

There is a consistent theme running through all these sections:

Prepare systematically

Ensure a sharp focus in your thinking

Opening moments are critical

Make positive use of non-verbal communication

Feed the recipient's expectations—let him know what's coming

Always remember the recipient's needs. Keep him well fed, appetisingly

A good transmitter merits a good receiver

Appendix 1

Report format for structural investigations

I am indebted to Mr Poul Beckmann of Ove Arup & Partners for permission to reproduce the following article

Writing a good report requires a skill which is rarely inherited and almost never imparted by the academic training of engineers. It is a skill which, like swimming, is not easily developed by study without practice, but which can be improved by paying attention to certain recommendations arising from experience.

Reports are written for a variety of purposes and each purpose imposes certain requirements of the presentation of the material. The following format is one which I have found useful for reports on structural investigations. Whilst I may have to modify it for certain reports, I find the discipline imposed by trying to follow it very helpful.

(1) Synopsis
One, or at the most two, pages of plain, succinct English, summarising the gist of the contents for the really busy top man, who will make his decision on the basis of this.

(2) List of contents
For the top man's assistant, so that he can quickly find the section dealing with a particular aspect of the report, when the top man asks about it.

(3) Brief
Who instructed us to do what, and when? Refer to the date and sender of the letter and quote the part of the letter which describes the expected extent of our work. Also quote any subsequent modification of our brief with date of letter or phone call.

This is useful to both parties: it helps us to make sure that we are answering all the right questions and it makes it clear to the reader what we were asked to do—or not do.

(4) Documents examined
List the documents made available to us and by whom (e.g. solicitor's letters, reports from others sent by solicitor, etc.) and describe those which we would have liked but did not get (e.g. structural drawings) with an

233

indication of why not. This will indicate to the reader from what basis of information our work was carried out and may justify why certain questions remain unanswered. It is also a check list to ensure that we have looked hard enough for information.

(5) Description of the structure

Just because you have bumped your head against every tie beam in the roof space, it doesn't follow that your reader has a clear picture of what you were investigating. Make it brief and pictorial, include a potted history of its construction and use and refer to diagrammatic drawings and perhaps photographs in the report. Avoid the catalogue style so beloved by some contributors to *The Structural Engineer* and the Arup *Newsletter*: your job is to convince the opposition, not bore them to death.

(6) Inspections

Who looked at what, how and when? It may be important to make it clear that an adequate number of inspections were made by adequately qualified people. The dates may be important in developing situations. Any limitations on the effectiveness of the inspections should be indicated (e.g. 'inadequate opening up, hence only superficial visual examination possible'). Follow with a concise description of what was seen on each occasion. If necessary refer to an appendix with a schedule of individual observations, but do *not* discuss inference or significance of observations in this section.

(7) Sampling and testing

Who took how many samples of what? When were they taken? When were they sent to whom for what kind of testing? Why? What were the results?

It may be important that samples are taken in the presence of representatives of the opposing side, hence name both the taker and the watcher if there is one. It is important that mechanical tests and chemical analyses are carried out by reputable independent laboratories, so name them. Give a brief précis of the results and refer to an appendix with photocopies of the laboratory certificates. Do not discuss yet.

(8) Design check

What relevant and necessary information was obtained from drawings, specification or original design calculations? Which necessary parameters have had to be assumed or deduced? What type of calculation was carried out and which criteria have the results been judged against? Summarize the findings and refer to an appendix containing the actual calculations. Do not draw any conclusions yet.

There may be occasions when it is advantageous to present the activities covered by Items 6, 7 and 8 in chronological order, even if it means that each item may be split into two or more parts, e.g. 'Initial inspection'—'General design check'—'Detailed inspection of . . . '—'Detailed stress analysis of . . . '—'Sampling and testing of . . . ', etc.

Up to this point the report has described facts which can be, or could have been, checked. What follows is interpretation and/or opinions which can more easily be challenged. The report will be safer to use and easier to defend in a possible conflict situation if 'facts' and 'fiction' are clearly separated.

234

(9) Significance of findings: (or 'Discussion')
As the heading indicates, this is the item under which you discuss the importance of each of the findings described under 6, 7 and 8 and, particularly, their relevance to the questions raised in the brief.

In the case of a failure investigation this is where you let your imagination loose; list all the causes that you or anybody else could reasonably think of and then compare each one, in turn, with the evidence. This way you should, with luck, be able to eliminate most of the hypotheses as being inconsistent with the facts.

(10) Conclusions
If Item 9 has been well written, a reasonably intelligent reader will by now have arrived at the correct conclusions unaided. This item need therefore only contain a brief paragraph stating in plain English the (by now obvious) answer to each of the question in the brief.

(11) Recommendations
A brief description of the course(s) of action which we recommend as the logical follow-up to the conclusions. Stick to the broad principles and describe them in clear, plain language, intelligible to the lay reader (e.g. the top man's assistant). Details, even if commissioned, should be banished to an appendix with a reference.

The many references to 'plain English' are prompted by the fact that we are so used to dealing in our daily lives with technical minutiae, which are most conveniently described in engineering jargon, that we forget that the end result is rather simple bits of hardware, which the layman can understand if it is described in plain words. It takes a little effort, but it can be done—and, in the case of reports, it is worth it!

Appendix 2

Marketing report

A possible sequence for putting a proposal to clients is:

1. Client's requirement
2. Our competence
3. Creative possibilities
4. Action required

This sequence has the following advantages:

1. Client's requirement: immediately, the reader recognises that his interests are our first concern. Feed them back to him, either in our words or preferably as a facsimile of his own statement.
2. The statement of 'our competence' ranks alongside the client's interests. References to similar previous work carries more weight than vaguer claims to competence.
3. The statement of 'creative possibilities' is one which implies, 'We *jointly* can do this *together*'. It leads naturally to:
4. Action required.

In presenting such a report, there will be need to precede it with an introduction. For a short report of, say, a couple of pages this may simply be a boxed statement immediately after the title, defining the purpose of the report.

For a longer report, there is the usual need, in addition to purpose, to tell the reader of the scope and shape of the report that is to come.

The section 'our competence' can grow to unwieldy proportions. There is a growing habit of including in such a section the curricula vitae of key people who might be assigned to the project—CVs which can grow to several prosaic pages each. These become extremely tedious in the main body of a report, and should be relegated to appendices. The main section of 'our competence' should be simple and straightforward: a statement of experience in corresponding projects.

There should be sufficient examples to help credibility, but not (in the main text) a long and potentially boring list. By all means, if you see fitting, offer the reader a wealth of evidence, but put this detail where he is not forced to read it as part of the main body.

Where it is appropriate to nominate individuals who would be concerned with the project, make their joint eminence and suitability clear in two or three well-phrased paragraphs of main text, and refer for detail to appendices.

Appendix 3

Standard structure—technical report

Below are listed typical headings for a technical report. Comments on each section are included in Chapter 11 of this book.

Abstract
>Contents
>>Introduction
Statement of problem
Acknowledgements
References
Literature survey
Experimental methods used
Problems encountered
Summary of results
Comments
Conclusions
Recommendations
Appendices